# Soil Erosion Issues in Agriculture

# Soil Erosion Issues in Agriculture

*Editor*

**Aleix Campo**

scitus academics

**Soil Erosion Issues in Agriculture**
Edited by **Aleix Campo**

ISBN: 978-1-68117-582-9
Library of Congress Control Number: 2016946437


SCITUS Academics LLC,
616, Corporate Way, Suite 2, 4766,
Valley Cottage, NY 10989

www.scitusacademics.com



**Notice**



Printed in United States of America on Acid Free Paper

# Preface

Soil is the earth's fragile skin that anchors all life on Earth. It is comprised of countless species that create a dynamic and complex ecosystem and is among the most precious resources to humans. Increased demand for agriculture commodities generates incentives to convert forests and grasslands to farm fields and pastures. The transition to agriculture from natural vegetation often cannot hold onto the soil and many of these plants, such as coffee, cotton, palm oil, soybean and wheat, can actually increase soil erosion beyond the soil's ability to maintain it. Soil erosion is a gradual process that occurs when the impact of water or wind detaches and removes soil particles, causing the soil to deteriorate. Soil deterioration and low water quality due to erosion and surface runoff have become severe problems worldwide. The problem may become so severe that the land can no longer be cultivated and must be abandoned. Many agricultural civilizations have declined due to land and natural resource mismanagement, and the history of such civilizations is a good reminder to protect our natural resources. Erosion is a serious problem for productive agricultural land and for water quality concerns.

*Soil Erosion Issues in Agriculture* deals with several aspects of soil erosion, focusing on its connection with the agricultural world.

Controlling the sediment must be an integral part of any soil management system to improve water and soil quality. Eroded topsoil can be transported by wind or water into streams and other waterways. The effects of soil erosion go beyond the loss of fertile land. It has led to increased pollution and sedimentation in streams and rivers, clogging these waterways and causing declines in fish and other species. And degraded lands are also often less able to hold onto water, which can worsen flooding. Sustainable land use can help to reduce the impacts of agriculture and livestock, preventing soil degradation and erosion and the loss of valuable land to desertification.

# Table of Contents

# Chapter 1

# Soil Erosion from Agriculture and Mining: A Threat to Tropical Stream Ecosystems

Karl M. Wantzen [1,*] and Jan H. Mol [2]

[1] River Culture (Fleuve et Patrimoine), Interdisciplinary Research Centre for Cities, Territories, Environment and Society (CNRS UMR CITERES), Université François Rabelais, Parc Grandmont, 37200 Tours, France

[2] Department of Biology, Center for Agricultural Research in Suriname (CELOS), Anton de Kom University of Suriname, Leysweg, Paramaribo, Suriname

## ABSTRACT

In tropical countries soil erosion is often increased due to high erodibility of geologically old and weathered soils; intensive rainfall; inappropriate soil management; removal of forest vegetation cover; and mining activities. Stream ecosystems draining agricultural or mining areas are often severely impacted by the high loads of eroded material entering the stream channel; increasing turbidity; covering instream habitat and affecting the riparian zone; and thereby modifying habitat and food web structures. The biodiversity is severely threatened by these negative effects as the aquatic and riparian fauna and flora are not adapted to cope with excessive rates of erosion and sedimentation. Eroded material may also be polluted by pesticides or heavy metals that have an aggravating effect on functions and ecosystem services. Loss of superficial material and deepening of erosion gullies impoverish the nutrient and carbon contents of the soils; and lower the water tables; causing a "lose-lose" situation for agricultural productivity and environmental integrity. Several examples show how to interrupt this vicious cycle by integrated catchment management and by combining "green" and "hard" engineering for habitat restoration. In this review; we summarize current findings on this issue from tropical countries with a focus on case studies from Suriname and Brazil.

## KEYWORDS

agricultural catchments; headwater stream; siltation; suspended sediment; turbidity; environmental impact; biodiversity

## 1. INTRODUCTION

Soil erosion caused by human land use is a widely known and intensively studied subject in agronomic sciences [1] and biogeochemistry [2]. The mobilisation and the transport of soil particles primarily causes losses of the fine and less dense particle fractions, including humus particles and clay, both of which are important carriers of soil nutrients and stabilizing agents for the physical properties of soils. The global areas of land area affected by erosion are estimated at 1094 million ha (Mha) by water erosion, of which 751 Mha is severely affected worldwide, and 549 Mha by wind erosion, of which 296 Mha is severely affected [3]. The economic losses by deterioration or complete loss of agricultural surfaces amount to billions of dollars [4].

Much less attention has been paid to the fate of these eroded particles when they have left the managed land. Our mindset is used to associate a brownish colour to the image of a "river", however, in many to most cases this colour does not derive from natural erosion processes in mountainous areas but rather relates to human activities in the catchment that increase erosion and sediment load in running water systems that drain these impacted areas. Rivers are transport systems, that carry a certain amount of solid substances. The sediment budget depends on the geological source material, e.g., the sediment-rich "white-water" rivers drain the geologically young Andes Mountains whereas clear- and black-water rivers that drain the weathered Precambrian Brazilian and Guiana Shields have a naturally low sediment load (terminology by Sioli [5]). Due to man-made increase of solid inputs into rivers, however, many river systems have to cope with unnatural amounts of solid discharge, especially in their headwater sections, and may become "overloaded" in the literal sense of the word. This overload is caused not only by agricultural erosion. Other human activities, such as mining for mineral resources at the soil surface, forestry, and badly planned road construction and/or drainage projects (both of which are often linked to agricultural activities) also may increase the solid discharge of streams and rivers beyond the natural dimensions. The effects on the biota are complex [6], but currently well recognized as one of the main anthropogenic threats to running water ecosystems in the world [7] including the Tropics [8]. The effects are also quite similar whatever the origin of the sediment overload. This is why this review deals with different aspects of sediment overload in running water systems on aquatic and riparian biota.

In tropical countries, the ecological setting in rivers and streams differs in a certain way from temperate systems [9,10,11]. Predictable spring floods deriving from snow melt, are well known in the temperate zone, but in the Tropics they are restricted to areas in the foothills of high mountains, e.g., in Ecuador [12]. Tropical rainfalls may provoke very short-termed and large changes in the discharge volume; increases up to 16 fold are reported from Latin America [13]. Wide areas in the Tropics have seasonal discharge patterns including extensive rainy periods once or twice a year and dry seasons between them. Each subsequent rainfall event over a rainy season increases the discharge. Consequently, the pulses of sediment entering the streams from erosion of gullies allow only very short intervals (if any) for the stream organisms to recover [14,15], and the accumulating effects of repeated single events are much stronger than their sum. Even during the dry season when lateral erosion gullies cease to deliver sediments and the stream discharge is reduced, the sand deposits that have covered the riparian zone during the rainy season gradually fall back into the stream channel due to bank slumping of the instable sediment deposits, prolonging the effects.

Many factors that enter into the Universal Soil Loss Equation result in non-linear increases in soil erosion in the Tropics [16]. For example, owing to the high erodibility of old soils in the Tropics (compared with the young ones of glaciated temperate areas), intense rainfall events with high energy impacts can be particularly erosive. The still widespread occurrence of inadequate soil management techniques may also increase erosion processes in the Tropics relative to the temperate zones. Subsurface drainage, a common source of fine sediments in temperate streams, is (still?) less important in tropical agriculture. Moreover, while in temperate areas the potential for agriculturally useable areas is asymptotically reaching the limits of growth, many countries still have a large potential of useable areas, and the land use change from native vegetation into agricultural and pasture areas is most intensive in the Tropics [17]. Generally, erosion problems occur during the first years of cultivation rather than during routine management.

The erosion impacts of soil surface mining are the same as in agriculture, *i.e.*, the additional sediments reach the stream via small lateral channels. The origin and type of sediments depend on the type of erosion. Gully erosion and superficial mining using water jets often mobilize the upper 1–10 m, which are mostly inorganic soils, whereas superficial sheet erosion in agricultural areas produces a larger proportion of humic substances, other organic substances and fine particles. However, there is a big difference in the governance of mining. In tropical countries, large mines of transnational companies have their environmental problems, including erosion from mine-waste dumps and roads, but they usually keep up with (international and local) environmental standards/legislation and the mines are mostly easily accessible and easily controlled. However, artisanal miners often ignore regulation about decantation of sediments mobilized for mining, if legislation exists at all, and law enforcement is hampered by the large number of remote and small-scale mining sites. The current gold rush in tropical South America was triggered in 1980 by the discovery of a 90,000 kg gold deposit in Serra Pelada, Amazonia, Brazil, and, in 1989, the number of individual miners in Amazonia was already estimated as high as one million [18]. Most of these miners are typically poor people driven by survival and with no subsistence alternatives [18,19]. The 400% increase of the price of gold over the last decade has further stimulated artisanal gold mining in the Amazon [20]. The miners are very mobile and illegally exploit secondary, alluvial surface gold deposits in small-scale mines in (former) stream valleys or they are involved in medium to large scale dredging operations on large rivers. They are extremely difficult to control and also wreak havoc with streams in nature reserves, even when these reserves are small and accessible (e.g., Brownsberg Park in Suriname) and thus theoretically easy to patrol. When easily extractable ores are depleted or legislation and law enforcement make mining too difficult they migrate to other regions and often cross borders into neighbouring countries. Thus the larger portion of the estimated 35,000 artisanal miners in Suriname is illegal immigrants from Brazil [18]. The mining method of hydraulic extraction of soil with high pressure water jets is basically artificial erosion at a very high rate. At the mine site, settling ponds that could prevent sediments from entering streams are not constructed and the riparian forest that could trap eroded sediments is removed. This results in a consortium of individual sources of increased sediment loads, which accumulate along the river. Thus the additive discharge of numerous "muddy" tributaries with gold mining in their catchments can even affect conditions in large rivers: for example on 5 February 2012 an unusually low Secchi disc transparency of 5–10 cm was measured during low flow conditions in the large Marowijne River in Suriname (catchment 68,700 $km^2$, discharge at outfall 1785 $m^3 s^{-1}$, pre-mining sediment discharge 18.9 ton $year^{-1} km^{-2}$), a clear-water river draining the Guiana Shield with a natural transparency of 100–200 cm (JM pers. observations). Mercury that is used by the miners to extract gold enters the stream and aquatic food web [21], is transported downstream with the suspended sediment [22], and continues to affect the ecosystem and human health for a long time after mining activities have ceased [18].

## 2. RESULTS AND DISCUSSION

In this review, we focus on describing the entry paths of eroded particles into the streams and rivers and their effects on the aquatic and riparian flora and fauna, and discuss prevention strategies in tropical countries.

### 2.1. Entry Paths of Eroded Particles into the Streams and Rivers

Once mobilized, eroded particles follow the hydraulic gradients and end up in the hydrographical network of rivers and streams. The origin of these particles can be manifold. Here, we focus on agricultural and mining processes.

#### 2.1.1. Agriculturally Caused Erosion, Including Drainage and Earth Road Construction

As a number of detailed studies on agricultural soil losses are published in this volume, we will only give a short summary of the processes that are relevant for the erosion process as such. We focus primarily on the transfer of eroded soil material into the streams and rivers, with a bias towards personal observations and studies in the Brazilian Cerrado savannah landscape. Most of these features, however, can be found in wide tropical areas.

The constant pressure for increased productivity, driven by growing market prices, leads to an intensification of agricultural use, with a range of negative effects for the ecological integrity of the landscape, such as erosion. The reformulation of the Brazilian Forest Code in December 2012 has strongly reduced the law enforcement against farmers who did not provide sufficient protected areas, and it also has allowed the use of permanently protected areas mentioned above, which may cause a step by step size reduction of these areas below the critical minimum size. In the State of Mato Grosso alone, the renewed Forest Code exempts 8 million hectares from reforestation (as it would have been necessary according to the previous version of the law), while another 6 million hectares will have to be reforested even now. Reinforcement for permanently protected areas will be exempted on 400,000 ha, while 500,000 ha still have to be recuperated [23]. The general recommendation by the advisory teams of the governmental agricultural consultancies is to use sandier soils for pasture, and clayey soils for agriculture. Pasture is also the prevailing land use for recently deforested areas along the agricultural expansion frontier in Southern Amazonia [17]. Very often cattle are allowed to move freely through very large areas. Drinking water is rarely provided for the animals, so that cattle compact soils on their trails in the riparian zone when looking for water. These cattle tracks are very often perpendicular to the contour lines; thus they are prime sites for initializing gully erosion. Specific habitats that are permanently protected by the Brazilian Forest Code [24], such as the riparian belt of 30–200 m, swamp forests, marshy campo wetlands, and steep hills, often become cut by cattle tracks [25]. All of these habitats are specifically sensitive to erosion. However, even in the more developed cattle ranches that keep the animals out of the sensitive riparian zone, erosion also occurs due to inadequate choice of exotic fodder species. Failure of seeding, or occurrence of pests (e.g., leaf-cutter ants) often results in bare soils sensitive to erosion. Several species of fodder plants grow in tufts that leave open space between them, which are preferred cattle tracks in the pasture landscape that serve as acceleration pathways for surface runoff. Lastly, we learned during field interviews that the degree of professional training and the general attitude of ranchers and farmers considering soil protection may be quite different. While cash crop farmers very often have a degree in agronomy, and have a very careful attitude towards soil conservation, many ranchers dispose over very large areas, where the occurrence of erosion is a minor problem, and erosion gullies only become fenced off when they turn into mortal traps for cattle.

Due to the use sequence "deforestation → cattle ranching → cash crops", we find in the Brazilian agricultural frontier region that the farmers planting cash crops often inherited erosion problems from the former owners (*i.e.*, Harding's "ghost of land use past" [26]). To mitigate the problems, contour ploughing, development of elevated contour lines, specific care for road construction and road drainage was undertaken. In the past 15 years, the percentage of direct planting (without or with minimal ploughing) has had considerable positive effects on soil quality [27] but a still unknown effect is the large-scale use of pesticides such as "roundup" that are used to kill off the interim plantation between the cash crops and to provide bare soil without tilling. Still, the erosion problem locally persists, especially if gullies reach the water table and continue growing even during the dry season in dendritic patterns (so-called "voçorocas" in Brazil).

An early study of erosion problems during the phase of most intensive land use change in Mato Grosso [28] found that most erosion features formed during the construction of earth roads in agricultural areas rather than during agricultural use itself. Earth roads allow accumulation and acceleration of surface water, which have an enormous destructive potential. In the long-term observation site near Jaciara, Brazil (16°02'30" S, 54°59'45" W), an erosion gully developed during a single rain event in ca. 1984 (Figure 1a, foreground) due to an earth road perpendicular to the contour lines. The farmers built a new parallel road, which caused the development of another gully (Figure 1a, middle ground), the ecological effects of which have been intensively studied [14,15,25,29,30]. A third parallel erosion gully had developed by 1994 due to cattle trampling (Figure 1a, middle ground). The gullies in agricultural areas of the Cerrado may be several kilometres long and as deep as 15 meters. They drain water from the surroundings (Figure 1 b). At its mouth the gully deposits large amounts of sediments during the rainy season, which are remobilized throughout the year, even during the dry season (Figure 1c). Two Wishmeier plots on deposits of the first gully from 1984 show how much sediment becomes mobilized, and that vegetation cover has a double impact on reducing the total amount of sediments and of keeping the average grain size of remobilized sediments small (Figure 2 a,b). After a while, the gully itself develops stream-like characteristics (Figure 1d), however the successions of the animal and plant communities remain in the early pioneer stages, as rainy season discharges destroy biological surfaces and "reset" the system [14]. In Suriname, similar >2 m deep erosion gullies also developed after heavy rains on earth roads in savannah vegetation; water turtles (*Rhinoclemys punctularia*) have been collected in these erosion gullies (JM pers. observations). DellaSala *et al.* [31] have pointed out the importance of roadless areas and the relatively intact ecosystems they maintain for the conservation of threatened freshwater biodiversity and ecosystem services to humans.

**Figure 1.** (**a**) Aerial view of a typical erosion gully in the Cerrado of Brazil. On the right, soy bean plantation showing bare soils after harvest, on the left, riparian vegetation surrounding a stream. In the foreground, an earth road and two erosion gullies of different age. The erosion gullies breach the riparian vegetation (capture date of photographs July 1995); (**b**) Active erosion gully in Figure (**a**). Deep lateral piping hollows drain groundwater from the surroundings, which flows off through the channel even during the dry season; (**c**) Stream receiving sediment from the active erosion gully in Figure (**b**) during the dry season. The gully delivers up to 60 metric tons of sediment per day during the rainy season,; (**d**) Pioneer vegetation developing on the bottom of the erosion gully in Figure 1**c** during the dry season. © Karl M. Wantzen.

Basically, every land owner is responsible for creating zones where this water can be captured and allowed to percolate, but the reality shows that most owners are mainly interested in getting rid of the problem by constructing water collecting channels, and sending the water to a lower-lying neighbour, resulting in a cascading increase in quantity and velocity of surface water. Eventually, streams and tributary gullies flowing into them are used as the sewers in the farmscape; thus the problem for the farmer has been solved, but the problems in the streams are just beginning.

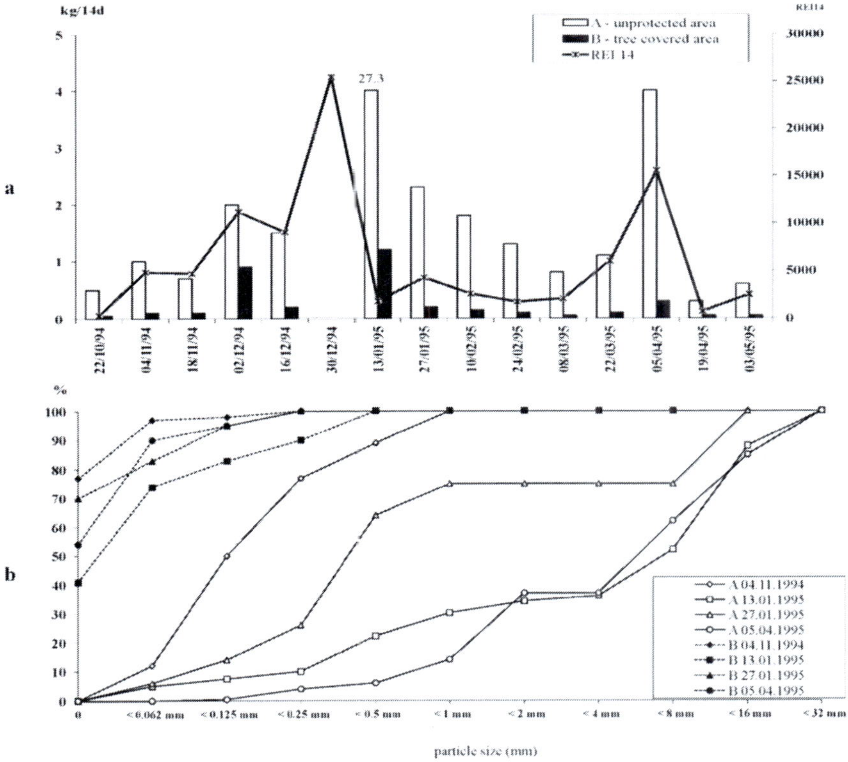


**Figure 2.** Mobilization of sediment on Wishmeier-plots (2 × 2 m) in the riparian zone of the stream near the mouth of the erosion gully shown in Figure 1c. (**a**) Amount of mobilised sediment (kg per 14day, bars, left axis, a maximum value of 27.3 kg was found on 13.1.1995 for a 4-week interval), REI 14 (RainEventIndex) was calculated as the sum of squares of the rainfall of the 14 days prior to measurement (line, right axis) (**b**) Granulometry of mobilised sediments on Wishmeier-plots from an unprotected site (A) and from a site covered by tree vegetation. Data from Karl M. Wantzen [29].

### 2.1.2. Erosion from Artisanal, Small-Scale Gold Mining Activities

Artisanal gold miners extract gold from easily accessible and exploitable, superficial, placer deposits in stream valleys using simple cheap technologies and then abandon the site when the reserves are depleted. Although "classical" geological exploration is typically not conducted, the trial-and-error discovery method is mostly restricted to areas with known gold-bearing geological formations, (e.g., the Greenstone Belt in eastern Suriname as opposed to western Suriname where primary gold ores do not occur and gold miners are absent). Erosion and siltation of streams are caused by both the method of gold extraction and the removal of the riparian forest along the stream. A mining site is prepared by first removing the riparian forest (Figure 3a,b) which leaves an approximately 200 m wide strip of bare soil along the stream exposed to erosion after heavy rain showers. Peterson and Heemskerk [32] estimated that 2300 $km^2$ of forest in Suriname would be cleared by artisanal miners in 2010; this may be a small fraction of the total area of Suriname rain forest (149,800 $km^2$) but the forest cleared by gold miners is all riparian forest (Figure 3b) and recovery is slow [32]. A recent estimate of mining related deforestation in Suriname based on Landsat images shows an increase in cleared forest from 85 $km^2$ in 2000 to 280 $km^2$ in 2008 [33]. Gold is mined with high-pressure water jets that remove the topsoil and "fluidize" the gold-bearing layer of sand and clay water (Figure 3c) followed by gravity separation of the heavy fraction of the slurry in a sluice box (Figure 3d) and discharge of the superfluous lighter sediments into old mine pits, adjacent to the forest or the stream. The heavy, gold-bearing fraction is amalgamated with mercury and the Au-Hg amalgamate is then separated from the undesirable mineral portion by panning, often at the stream margin. Finally the gold is recovered from the amalgamate by burning in simple pans which removes most of the volatile Hg (with Hg lost to the atmosphere and then after condensation deposited in nearby terrestrial and aquatic ecosystems). The sediment discharge of a mining-impacted stream in Suriname was estimated at 310 tonnes year $km^2$ (of which 95.6% was produced by the 2.5 $km^2$ mine site) as compared to a sediment discharge of 13 tonnes year $km^2$ in an undisturbed neighbouring stream (Table 1).

A bird's eye view of an active gold mining site shows a stream valley cleared from forest and with the stream itself replaced by a series of water-filled mine pits with colours ranging from bright green (algae) to blue to light brown. Downstream from the mining site the stream is still covered by the closed forest canopy as in a pristine catchment, but when studied closely the diversity of instream habitat and fish assemblage (Figure 3e and below) are low when compared to a stream in an undisturbed catchment (Table 1; [34]). When the gold deposits are depleted and miners migrate to another area they leave behind a legacy of environmental problems. Forest recovery following mining is slow and qualitatively inferior compared to regeneration following other land uses: large parts of mined areas remain bare ground, grass, and pits with standing water even four years after the miners have left the site (Figure 3f). Fine sediments contaminated with mercury are present at the former mine site (on land and in water-filled mine pits) and downstream of the site in the stream.

**Figure 3.** Artisanal gold mining, soil erosion and stream sedimentation in Suriname; (**a**) In 2005, the riparian forest was cleared at a recently developed gold mining site along Maykabuka Creek, Gros Rosebel Area (05°04′45″N, 55°16′9″W), In 2001, the then pristine Maykabuka catchment

still had the forest canopy closing over the stream and the site in the photograph was used as a control in a 1994–2001 study of the effects of gold mining related erosion on instream habitat and fish community [34], (capture date of photograph 8 December 2006); (**b**) Satellite image showing extensive clearance of riparian forest associated with small-scale gold mining in the catchments of the Merian (05°06′ N, 54°31′ W) and Tumatu rainforest streams, Marowijne River Basin, Suriname, September 2010, Source Quickbird; (**c**) High pressure water is used to remove the topsoil and then the gold-bearing layer of sand and clay, (capture date photograph of 7 May 2008); (**d**) In a sluice box heavy particles and gold are gravity separated from the superfluous lighter particles in the slurry which then are discharged into an abandoned mining pit in the adjacent forest or stream, (capture date of photograph 7 May 2008); (**e**) Downstream a mining site in the Gros Rosebel Area, a shallow dry-season pool in a turbid rainforest stream with intact riparian forest and closed canopy cover (4°30′ N, 55°20′ W) has low dissolved oxygen (1.3 mg/L) and an unusually low fish diversity virtually monopolized by the auchenipterid catfish *Trachelyopterus galeatus*, (capture date of photograph 5 November 2008); (**f**) An abandoned mine site in the Merian Creek catchment (05°06′44″ N, 54°31′16″ W) shows slow riparian forest regeneration, abundant growth of grasses and a shallow, turbid, sunlit stream, (capture date of photograph 26 November 2011). ©Jan H. Mol.

## 2.2. Effects of Increased Sediment Load on Habitats and Biota in and Near Streams

Sediment, derived from soil erosion from agricultural sites or from surface mining entering a stream in large quantities, changes many environmental conditions at the same time.

### 2.2.1. Effects on Stream Habitat Structures

Surfaces of natural and artificial substrates protruding into the current become scoured by the suspended sediment that acts like a blaster used for abrasion of unwanted graffiti. Consequently, epilithic algae covers (biofilms) are brushed off. It could be shown experimentally that single rain events can diminish these biotic layers drastically and therefore it was concluded that scouring effects limit food for grazing organisms, e.g., macroinvertebrates [30] and loricariid catfish [35].

The particles that remain in suspension increase the turbidity of the water, reduce light transmission [36], and thus affect the aquatic food web, both from the bottom up through reduced photosynthesis and top-down by limiting visual foraging efficiency of many fish [37]. The population of visually oriented animals decreases in streams impacted by anthropogenic sediment [34]. Many of these are predators, which play an important role in the structure of the entire biotic community (*i.e.*, top-down effects). Plants depend on light. Other than temperate streams that are mostly driven by allochtonous organic leaf litter falling from the riparian trees, tropical streams have an additional food source coming from autochtonous algal production [38]. Even though there are limited light spots or lit periods on the stream bottom, these are sufficiently strong to support photosynthesis. Stable isotope studies have shown that in many tropical streams, algal production is very important [39,40]. Increased turbidity of these naturally clear streams therefore changes the baseline of the entire food web (*i.e.*, bottom-up effects).

Lastly, deposited sediment *covers* habitats and food sources. Sediment settles in low-current zones of the streams (e.g., pools), which are important deposition zones for organic matter. Woody debris, leaf litter and fine organic particles which represent an important source of food or shelter for most aquatic animals become completely covered by sediment [34]. Apart from being unapproachable by animals, the sediment-covered organic materials start decomposing in an anaerobic pattern, which has severe negative impacts on water quality, such as depletion of dissolved oxygen and, in mining areas, methylation of mercury (which allows for accumulation and biomagnification of mercury via the food chain). Even if the riparian vegetation still appears to be nearly natural, the increased solid discharge causes the loss of most of the biodiversity in the stream channel (e.g., downstream of a mining site [34]).

### 2.2.2. Effects on Stream Plants and Animals

As shown above, the physical structure of habitats and food sources become more limited or is destroyed. Apart from these indirect impacts, aquatic biota also suffer direct impacts from excessive loads of suspended or bottom-transported sediments. The mechanisms by which and where the suspended solids interfere with the aquatic organisms are generally known [6,41,42].

Aquatic macrophytes often colonize low-energy habitats, where sediment settles on their leaves and blocks photosynthesis. Increasingly thick sediment layers cause anoxia and reduce the possibility of nutrient uptake by the roots. The branched structure of macrophytes makes them ideal sediment traps; indeed, sediment accrual is a growth strategy of many water plants [43]. When sediment loads increase, however, this strategy causes fast burial of the macrophytes [44]. Water plants and roots from terrestrial plants protruding into the high current zone of the stream bed become sheared off. This has especially severe effects on the entire ecosystem whenever and wherever these plants have an important structuring role by providing habitat for most of the other biota. In tropical rivers, plants of the family Podostemacea live in the shallow, fast flowing water of rapids where they are estimated to harbour up to 50% of the animal and plant biomass on just 1% of the stream bed [45]. Many fish that dwell in rapids, feed on invertebrates living in the Podostemacea vegetation or on the plants themselves [46] and find shelter from predators and from the river current between the submerged leaves. Odinetz Collart *et al.* [47] suggest that the Podostemacea vegetation is sensitive to smothering by deposited sediment from gold mining.

Filtering organs, gills and sensitive body surfaces of animals become clogged or abraded. As most aquatic animals breathe either via gills or sensitive membranes on their body surface, this effect is deleterious for the entire fauna. Siltation quickly eradicates large mussels, as these have a limited potential to move away from the impact, and are especially sensitive to clogging of their filter-feeding and breathing organs [48]. In a survey of aquatic macroinvertebrates above and below the confluence of a stream with an erosion gully, practically all taxonomic groups showed dramatic decreases with siltation impacts, with the exception of animals that are able to dig in sand and/or to breathe at the water surface [14]. Experiments with stoneflies (genus *Anacroneuria*) in artificial flumes have shown that drift was significantly increased by increasing sand concentrations (MANOVA, $p < 0.0001$, $n = 150$); [29]. Benthic invertebrates perceive the increased sand load in the water within less than a second and start to search for downstream habitats, which may, however, already be sand-covered. A study trying to establish a bio-indicator system for measuring the degree of severity of sand deposits using standardized artificial substrates [49] revealed that practically all taxonomic groups of invertebrates suffer from siltation impacts alike. There is a gradation of organisms that prefer substrata with solid surfaces such as stoneflies (Plecoptera), which are the most severely reduced, compared to organisms that can dig into softer substrates such as non-biting chironomid midges, which may recover more quickly from the siltation impact (Figure 4). The life-cycle length and the reproductive strategy of the organisms are very important in determining their sensitivity to changes in their physical habitat by processes like siltation. The longer-lived stoneflies became completely eradicated from sites downstream from an erosion gully, while the short-lived chironomids (life cycle length of less than three weeks) could still colonize the artificial substrates in our study (see [14] for a detailed analysis).

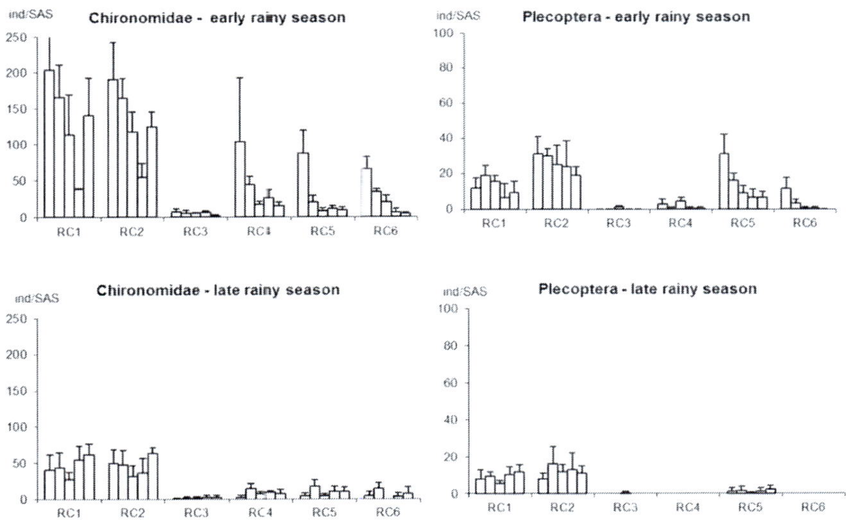


**Figure 4.** Accumulating effects on the abundance aquatic insects (Chironomidae and Plecoptera, individuals per standardised artificial substrate, averages plus standard deviation) from the beginning (22 October 1994–16 December 1994) to the late (8 March 1995–3 May 1995) rainy season (Each pillar represents a 14 day interval) in a Cerrado stream. RC1, RC2: reference sites above the confluence with the erosion gully (RC3), RC4-RC6: impact sites with growing distance below the confluence. Data from Karl M. Wantzen [29], see [14,49] for site and method description.

Impacts from increased anthropogenic loads of suspended and deposited sediment on fish are best studied in temperate salmonid fish from the Northern Hemisphere [50]. The effects of sediment on fish are complex and often include cumulative or synergistic effects with other stressors that are difficult to capture in laboratory studies. Effects can be lethal (individual fish are killed) or sublethal (e.g., tissue damage or alteration of physiology resulting in reduced growth), or behavioural. The magnitude of the impact depends on the concentration of suspended sediment, type of sediment (particle size, angularity), duration of exposure (sort pulse *vs.* chronic), natural background levels (e.g., high flow conditions *vs.* low-flow conditions, white-water *vs.* clear-water rivers, position in the river continuum as in headwaters *vs.* estuary), species (clear-water species, like salmonids, that are not normally exposed to high sediment levels are most sensitive) and life stage (early life stages such as eggs and larvae are often most sensitive) [50,51]. Turbidity reduces plant biomass and food availability, as well as the visibility of pelagic food, and also reduces predation risk from visually-oriented piscivores and fishing birds; particles clog gillrakers and gill filaments and reduces benthic food availability [52]. Salmonids are known to avoid turbid water [50] and thus anthropogenically enhanced suspended sediment concentrations can affect their upriver spawning migrations and headwater ecosystems [53]; such information is currently lacking for tropical migratory fish species [54]. The deposited sediment covers habitat structures (leaf litter, submerged root masses, woody debris, gravel beds, macrophyte stands) that fish use for shelter or reproduction ([34]; Table 1). The interstitial pore space of coarser (pebble) sediment in mountainous regions is a very important habitat for early life stages of fish (eggs, larvae) that can be lost by siltation. Many fish of temperate zones such as salmon and lampreys spend their egg and early life phases in this pore space. Siltation causes clogging and reduction of habitat quality (e.g., reduced flow, reduced oxygen concentration, *etc.*), thus preventing fry emergence or making the gravel bed uninhabitable [55]. Human-induced change in optic conditions in water with increased suspended sediment degrades the sensory environment of fish and thus negatively affects visual communication [56], important in schooling [57], territoriality, and courtship [58]. Evolutionary consequences of high turbidity (associated with eutrophication) were revealed in Lake Victoria cichlids showing decreased intensity in male colour, decreased colour-mediated sexual selection, and a decreased number of coexisting colour morphs (*i.e.*, diversity) in turbid water [59]. Both richness and diversity of fish assemblages in Brazilian Cerrado streams were negatively influenced by siltation impacts caused by agriculture [60]. In the stream systems studied by Wantzen [14], several headwaters originally harbouring a rich fish fauna became entirely fishless as a consequence of siltation. In Suriname, two super-endemic catfish species (distribution restricted to a single mountain creek) are threatened with extinction by on-going gold mining and a proposed bauxite mining project in Nassau Mountains [61]. Mol and Ouboter [34] studied the instream habitat and fish assemblage of a small lowland rainforest stream in Suriname in a reach immediately downstream a gold mining site (Table 1). Whereas elevated turbidity associated with short pulses of sediment delivery to streams (e.g., after heavy rains flush sediments from earth roads in a tributary) apparently can offer temporary protection to prey fishes from visually-oriented piscivores (J. Mol pers. observations), the mining-impacted stream had chronically (1994–2001) high suspended sediment concentrations, high turbidity, low Secchi disc visibility, a thick layer of fine sediment covering bottom substrate, and low substrate diversity compared with an undisturbed control stream in a neighbouring catchment (Table 1). The fish of the turbid, mining-impacted stream were silvery or unpigmented as compared to fish from an undisturbed stream (Figure 5). Although the mining-impacted stream was slightly larger than the pristine control stream it had a lower number of fish species and lower fish diversity compared to the undisturbed stream (Table 1). The mining impacted stream had few visually-orienting fishes such as cichlids, callichthyid catfish (mainly *Corydoras* spp.) and eryrthrinid piscivores (mainly *Hoplias* spp.), but abundant fish that communicate and feed with olfactory and tactile (auchenipterid catfish), electric (gymnotiform knife fish) and lateral line (Gasteropelecidae or hatchet fish) senses (Table 1). The mining-impacted stream also had few juvenile fish and few large food fish (erythrinids, cichlids) compared to the pristine stream (Table 1). Lujan *et al.* [62] also found large, economically important fish (Pimelodidae and Serrasalmidae) absent from a mining impacted river in the foothills of the Andes Mountains. There are very few studies on effects of mining on tropical fish outside the Neotropics, but Moyle & Leidy [63] suggest that "in Sri Lanka, sedimentation from logging practices on steep rain-forest slopes and streamside mining has been a major contributor to the decline of the endemic fishes".

**Table 1.** Downstream impact of gold-mining related erosion on instream habitat and fish community structure of a small rainforest stream in Suriname in the period 1994–2001, data from Mol and Ouboter [34].

| | Undisturbed Stream | Mining-Impacted Stream |
|---|---|---|
| | Maykabuka Creek | Mamanari Creek |
| Monthly flow ($m^3 s^{-1}$) | 0.54–1.65 | 0.92–3.92 |
| Riparian rain forest | Undisturbed | Undisturbed |
| Total suspended solids ($mg L^{-1}$) | 19.0–28.9 | 318–2469 |
| Turbidity (NTU) | 28.2–31.1 | 424–2874 |
| Secchi disc visibility (cm) | >50 | <10 |
| Sediment yield (tonnes $year^{-1} km^{-2}$) | 13 | 310 (of which 95.6% produced by the gold mine) |
| Thickness of layer of fine sediment on the streambed (cm) | 0 | 12.8 (runs)–33.2 (pools) (maximum 57 cm) |
| Substrate diversity (Shannon-Wiener index) | 1.57 | 0.70 |
| Number of fish species | 68 | 56 |
| Fish diversity (Shannon-Wiener index) | 3.19–3.39 | 2.60–2.70 |
| Erythrinidae (% of total number of fishes caught) | 3.48 | 0.41 |
| Gasteropelecidae (%) | 4.81 | 18.61 |
| Gymnotiformes (%) | 3.54 | 12.62 |
| Auchenipteridae (%) | 1.20 | 3.33 |
| Callichthyidae (%) | 14.65 | 0.62 |
| Cichlidae (%) | 5.98 | 0.58 |
| Juvenile fishes (%) | 57.3 | 14.1 |
| Food fishes (% of total fish biomass) | 33.2 | 13.4 |

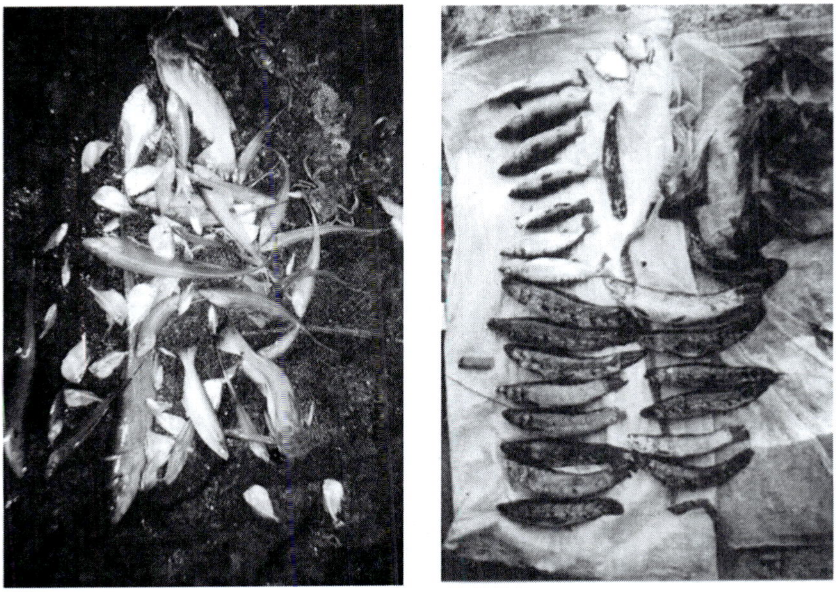

a b

**Figure 5.** Impact of small-scale gold mining on fish communities of rainforest streams as revealed by fish collections from two small rainforest streams in Suriname; (**a**) Surface feeding hatchet fishes and nocturnal electric knife fishes and catfishes dominate the catch in a mining-impacted stream; note the unpigmented or silvery colours of the fishes; (**b**) Large diurnal piscivores (*Hoplias* spp.) and brightly coloured fishes dominate the catch in an undisturbed neighbouring stream. See Mol and Ouboter [34] for details. ©Jan H. Mol (capture date of photographs April 2001).

## 2.3. Effects on Riparian Habitat Structures

Riparian forest buffer zones provide important protection of streams from surface runoff. The riparian forest also provides allochthonous food and habitat structure, such as woody debris and leaf litter to the stream ecosystem. Thus, the assemblage of fish is usually related to riparian forest cover [64,65]. In agricultural areas, erosion gullies may break through gallery forests and destroy their structures (see Figure 1a). These confluence zones are hardly colonized at all by metazoans (see RC3 in Figure 4) because a steady or pulse-like inflow of sediment-charged water abrades all biologically colonized surfaces, removing biofilms and larger organisms. The confluence site is no longer protected by riparian vegetation; therefore it is a starting point for further bank erosion. On the bank of the stream opposite the mouth of the gully, sediment deposits up to several meters high may accumulate, depending on the energy of the water flow during the last discharge event in the erosion gully. These sediment deposits are then gradually undercut by the water and slump into the stream channel, to be transported downstream and eventually redeposited. As a consequence of loss of bank stabilization the stream bank begins to widen and may become braided [25,66]. In consequence, streams become completely altered by siltation. In a first phase, dams of organic material may develop, forming small ponds. As soon as these soft biological structures are sheared off, the remaining tree trunks are reduced to single boles and the braiding can widen the stream channel from several meters to hundreds of meters (Figure 6).

At gold mining sites the riparian forest is completely cleared in a ≥100 m wide strip along the stream [32]. The riparian forest downstream of the mining site may be impacted by the deposition of fine sediments during high-flow flood pulses (Figure 7).

## 2.4. Effects on Riparian Animals and Plants

Deposited sediment layers cause anoxia and die-back of the riparian vegetation, whereas erosion initiating from the confluences of erosion gullies also undermines bank-stabilizing trees. In the first phase, the increase of retentive structures (dead trees) increases the residence time of organic matter (twigs and leaves) so that the number of decomposing organisms may be locally increased. The same is true for scavenging organisms and predators that can switch to the scavenging feeding mode. In the second phase (severe braiding), biodiversity is strongly reduced, only a few organisms (most of which have drifted down from the above stream sections) are found. Both in animals and plants, the ecological strategy switches from K-selected organisms (longer-lived, larger organisms that invest more energy in their offspring) towards r-selected species that produce large numbers of offspring in order to compensate eventual losses in these harsh and quickly changing environments [25,29,66].

**Figure 6.** Die-back of riparian vegetation and stream braiding due to excessive sediment deposition originating from agricultural erosion. See Wantzen [66] for details.© K.M. Wantzen, capture date of photograph July 1995.

## 3. POSSIBLE ACTIONS

The first step towards erosion prevention is monitoring of the extent and location of erosion. Satellite imagery can help to efficiently monitor turbidity levels in large rivers [67], land use and riparian deforestation along streams (e.g., Figure 3b). Second comes management of the erosion problem by developing policies, legislation, communication, and erosion-prevention techniques, all within a watershed perspective [8] and based on monitoring data, stakeholder-participation and scientific knowledge. The use of effective soil conservation techniques in agricultural areas situated above the riparian zone is of paramount importance. The most important action is to avoid the development of large bare or sealed surfaces by facilitating the infiltration and percolation of rainwater at the place it comes down. In addition to the well-known soil-protective measures in the fields (contour ploughing, direct plantation, *etc.*), this can be performed by creating or maintaining vegetated buffer strips of interfluvial vegetation with an additional grass strip between the natural vegetation and the pasture [25,66]. In the highly erosion-sensitive areas, the whole stream valley and its vegetation sequence have to be considered for conservation. Access of livestock and dirt road construction should be completely excluded from the waterlogged or otherwise erosion-prone areas of the valley in order to avoid trampling paths that facilitate the development of surface runoff. The maintenance of both buffer strips in the transition zone between land use and the riparian zone, and the breadth of the protected riparian zone require political and legal support [25]. Legal enforcement is urgently needed here.

Apart from prevention, several restoration methods are available to restore environmental quality of streams that have been impacted by excessive sediment loads. Various techniques have been proposed to restore erosion gullies [67]; however, few of them have been effectively tested. The paramount problem is to permanently stop the movement of bed sediments, continued deepening of the gully bottom, and subsurface extension of the erosion gully (dendritic piping). First, direct surface runoff into the gully needs to be prevented and percolation of surface water near the gully needs to be reduced (to prevent further piping). Buffer zones can significantly contribute to this task as they keep road construction and cattle out of sensitive riparian zones where most erosion gullies initiate. The second step in restoration is the development of an environment suitable for plants to recolonize the gully. This can be accomplished by dam construction inside the gullies and stabilization of the bottom of the gullies and their surrounding vegetation. Very often, erosion has removed the topsoil including the seed bank, soil nutrients, humus, and essential clay-soil aggregates. The restoration of this soil layer may be very expensive. Therefore in most cases it is avoided, and natural succession of plants is retarded. However, unusual measures such as fish pond construction in erosion gullies may have considerable success [25]. Active planting of plants is a third step in the restoration (see plant list for the Brazilian Cerrado in [66]). Special care has to be taken to avoid plantation of non-native species that may invade other areas and outcompete regional plant species. Once established, the vegetation may be quickly eradicated by periods of drought or high discharge. Therefore a two-step strategy with a rapid stabilisation of the soils and sediments by quickly establishing pioneer plants (or even sugarcane grids) followed by introduction of woody plants is recommended [66]. Once the problems caused by erosion are under control, classical stream restoration measures can be used. However, natural succession should be preferred where possible. Deeply incised erosion gullies and stream channels may require levelling of the bank structures by heavy (and expensive) machinery.

**Figure 7.** Fine white sediment originating from upstream gold mining is deposited in the riparian forest a ong Mamanari Creek, Suriname, after inundation of the floodplain forest during a high-flow event. See Mol and Ouboter [34] for details. © Jan H. Mol (capture date of photograph 12 April 2001).

To reduce impacts from gold mining, the Inter-American Development Bank recommends to (i) coordinate land-use planning across natural resources and resource use sectors (ii) tailor gold mining regulations and standards to background environmental and land use, (iii) support uptake of mitigation technologies to reduce sediment and mercury loading, (iv) encourage small operators to form collectives to normalize interactions and increase visibility, and to (v) internalize the environmental costs of impaired water quality and biodiversity loss [68].

## 4. CONCLUSIONS

Poor land use practices, including the removal of (riparian) forest cover, have caused increased rates of erosion in the Tropics. Land-based agriculture and gold mining in the headwater areas of catchments have resulted in chronic increases in the sediment load of streams. The increased suspended sediment load reduces light penetration and photosynthesis in the water and deposited sediment decreases instream habitat diversity by smothering aquatic macrophyte stands, clogging of gravels and hiding spaces, and covering of leaf litter, woody debris and other substrata. The stream ecosystem is affected both by direct impacts on the biota and by indirect effects via the food chain through reduced photosynthesis (bottom up) and reduced efficiency of visually-orienting predators (top down). As a result the structure (diversity) and functioning of the tropical stream communities have changed as is shown by changes in macroinvertebrate and fish assemblages. Nature reserves have been affected by mining and agriculture both within and upstream of the reserves. A consequence of the (longitudinal) hydrological connectivity of stream ecosystems is that ecosystems both up- and downstream of the point of sediment inflow can be affected by anthropogenic erosion disturbance (for example, turbidity inhibits upstream migration of fish and increases downstream accumulation of sediment). The accumulated sediment and associated contaminants are expected to leave a legacy of environmental problems long after the causes of erosion are corrected. Standards for turbidity and suspended solids should be set as some multiple of background levels upstream of a proposed activity [51]. The duration of exposure to elevated sediment levels should be included in the standards, as the impact of short pulses on the stream ecosystem is different from that of chronic exposure [41]. Monitoring data are the base of management, but they are often lacking in the Tropics [8]; however, turbidity levels in large rivers [69], land use practices, and clearance of riparian forest (Figure 3b) can be monitored efficiently using satellite imagery. A watershed perspective should be taken in addressing erosion-related problems in tropical streams [8], taking into account the legacy of past land use and cumulative effects from other stressors.

## ACKNOWLEDGMENTS

Paper # 1 of the UNESCO Chair "River Culture" and # 3 of the CARBIOCIAL project [70]. This paper resulted from discussions during the first ATIPA (Actions de formation Transfrontalières pour l'Intégration du Plateau des Guyanes en Amazonie) workshop in Cayenne, French Guyana, February 2013.

## REFERENCES

1. Lal, R. Soils and food sufficiency. A review. *Agron. Sustain. Dev.* **2009**, *29*, 113–133.

2. Quinton, J.N.; Govers, G.; Van Oost, K.; Bardgett, R.D. The impact of agricultural soil erosion on biogeochemical cycling. *Nature Geosci.* **2010**, *3*, 311–314.

3. Lal, R. Soil erosion and the global carbon budget. *Environ. Inter.* **2003**, *29*, 437–450.

4. Pimentel, D.; Harvey, C.; Resosudarmo, P.; Sinclair, K.; Kurz, D.; McNair, M.; Christ, S.; Shrpirtz, L.; Fitton, L.; Saffouri, R.; *et al*. Environmental and economic costs of soil erosion and conservation benefits. *Science* **1995**, *267*, 1117–1123.

5. Sioli, H. Das Wasser im Amazonasgebiet. *Forsch. Fortschr.* **1950**, *26*, 274–280.

6. Ryan, P.A. Environmental effects of sediments on New Zealand streams: A review. *New Zealand J. Mar. Freshwater Res.* **1991**, *25*, 207–221.

7. Malmqvist, B.; Rundle, S. Threats to the running water ecosystems of the world. *Environ. Conserv.* **2002**, *29*, 134–153.

8. Castello, L.; McGrath, D.G.; Hess, L.L.; Coe, M.T.; Lefebre, P.A.; Petry, P.; Macedo, M.N.; Reno, V.F.; Arantes, C.C. The vulnerability of Amazon freshwater ecosystems. *Conserv. Lett.* **2013**, *6*, 217–229.

9. Boulton, A.J.; Boyero, L.; Covich, A.P.; Dobson, M.; Lake, P.S.; Pearson, R.G. Are Tropical Streams Ecologically Different from Temperate Streams? In *Tropical Stream Ecology*; Dudgeon, D., Ed.; Elsevier: Amsterdam, the Netherlands, 2008; pp. 257–284.

10. Wantzen, K.M.; Junk, W.J. The Importance of Stream-Wetland-Systems for Biodiversity: A Tropical Perspective. In*Biodiversity in Wetlands: Assessment, Function and Conservation*; Gopal, B., Junk, W.J., Davies, J.A., Eds.; Backhuys: Leiden, the Netherlands, 2000; pp. 11–34.

11. Wantzen, K.M.; Mathooko, J.; Yule, C.; Pringle, C.M. Organic Matter Processing in Tropical Streams. In *Tropical Stream Ecology*; Dudgeon, D., Ed.; Elsevier: Amsterdam, the Netherlands, 2008; pp. 43–64.

12. Jacobsen, D.; Encalada, A. The macroinvertebrate fauna of Ecuadorian highland streams in the wet and dry season.*Arch. Hydrobiol.* **1998**, *142*, 53–70.

13. Lewis, W.M., Jr.; Hamilton, S.K.; Saunders, J.F., III. Rivers of Northern South America. In *River and Stream Ecosystems*; Cushing, C.E., Cummins, K.W., Minshall, G.W., Eds.; Elsevier: New York, NY, USA, 1995; pp. 219–256.

14. Wantzen, K.M. Physical pollution: Effects of gully erosion in a tropical clear-water stream. *Aquat. Conserv.* **2006**, *16*, 733–749.

15. Wantzen, K.M. Cerrado Streams—Characteristics of a threatened freshwater ecosystem type on the tertiary shields of South America. *Amazoniana* **2003**, *17*, 485–502.

16. Johnson, D.L.; Lewis, L.A. *Land Degradation: Creation and Destruction*; Blackwell: London, UK, 1995; pp. 1–335.

17. Nepstad, D.; McGrath, D.; Alencar, A.; Barros, A.C.; Carvalho, G.; Santilli, M.; Vera Diaz, M.C. Frontier governance in Amazonia. *Science* **2009**, *295*, 629–631.

18. Veiga, M.M.; Hinton, J.J. Abandoned artisanal gold mines in the Brazilian Amazon: A legacy of mercury pollution.*Nat. Resour. For.* **2002**, *26*, 13–24.

19. Heemskerk, M. Livelihood decision making and environmental degradation: Small-scale gold mining in the Suriname Amazon. *Soc. Nat. Resour.* **2002**, *15*, 327–344.

20. Swenson, J.J.; Carter, C.E.; Domec, J.C.; Delgado, C.I. Gold mining in the Peruvian Amazon: Global prices, deforestation, and mercury imports. *PLoS One* **2011**, *6*.

21. Ouboter, P.E.; Landburg, G.; Quik, J.; Mol, J.; van der Lugt, F. Mercury levels in pristine and gold mining impacted aquatic ecosystems of Suriname, South America. *Ambio* **2012**, *41*, 873–882.

22. Balogh, S.J.; Meyer, M.L.; Johnson, D.K. Transport of mercury in three contrasting river basins. *Environ. Sci. Technol.* **1998**, *32*, 456–462.

23. Secretaria de Assuntos Estratégicos (SAE) Impacto da revisão do Código Florestal: como viabilizar o grande desafio adiante? Internet report. Available online: http://www.sae.gov.br/site/?p=15735 (accessed on 24 September 2013).

24. Silva, J.A.A.; Nobre, A.D.; Manzatto, C.V.; Joly, C.A.; Rodrigues, R.R.; Skorupa, L.A.; Nobre, C.A.; Ahrens, S.; May, P.H.; Sá, T.D.A.; *et al. O Código Florestal e a Ciência: Contribuição para o Diálogo*; Sociedade Brasileira para o Progresso da Ciência & Academia Brasileira de Ciências: São Paulo, Brazil, 2011; pp. 1–124.

25. Wantzen, K.M.; Siqueira, A.; Nunes da Cunha, C.; Sa, M.F.P. Stream-valley systems of the Brazilian Cerrado: Impact assessment and conservation scheme. *Aquat. Conserv.* **2006**, *16*, 713–732.

26. Harding, J.S.; Benfield, E.F.; Bolstad, P.V.; Helfman, G.S.; Jones, E.B.D., III. Stream biodiversity: The ghost of land use past. *Proc. Natl. Acad. Sci. USA* **1998**, *95*, 14843–14847.

27. Batlle-Bayer, L. Changes in organic carbon stocks upon land use conversion in the Brazilian Cerrado: A review. *Agric. Ecosyst. Environ.* **2010**, *137*, 47–58.

28. Couto, E.G. *O uso da terra e o garimpo na bacia do rio Sao Lourenco, Mato Grosso: Reflexos No Ambiente*; FEMA/UFMT-CCA: Cuiabá, Mato Grosso, Brazil, 1990; p. 206.

29. Wantzen, K.M. Influence of Man-Made Siltation on Habitat Structure and Biotic Communities of Cerrado Streams of Mato Grosso. Ph.D. Thesis, Herbert Utz Verlag, Munich, 1997; p. 186.

30. Wantzen, K.M. Effects of siltation on benthic communities in clear water streams in Mato Grosso, Brazil. *Verh. Int. Ver. Limnol.* **1998**, *26*, 1155–1159.

31. DellaSala, D.A.; Karr, J.R.; Olson, D.M. Roadless areas and clean water. *J. Soil Water Conserv.* **2011**, *66*, 78–84.

32. Peterson, G.D.; Heemskerk, M. Deforestation and forest regeneration following small-scale gold mining in the Amazon: The case of Suriname. *Environ. Conserv.* **2001**, *28*, 117–126.

33. Centrum voor Landbouwkundig Onderzoek in Suriname (CELOS). Assessments of the extend of land categories, land uses and changes for the Second National Communication—AFOLU GHC Inventory 2011. CELOS, Natural Resources and Environmental Assessment: Paramarib, Suriname, 2011; unpublished work.

34. Mol, J.H.; Ouboter, P.E. Downstream effects of erosion from small-scale gold mining on the instream habitat and fish community of a small neotropical rainforest stream. *Conserv. Biol.* **2004**, *18*, 201–214.

35. Power, M.E. The importance of sediment in the grazing ecology and size class interactions of the armored catfish,*Ancistrus spinosus*. *Environ. Biol. Fish.* **1984**, *10*, 173–181.

36. Davies-Colley, R.J.; Smith, D.G. Turbidity, suspended sediment, and water clarity: A review. *J. Amer. Water Resources Ass.* **2001**, *37*, 1085–1101.

37. Aksnes, D.L.; Nejstgaard, J.; Sœdberg, E.; Sørnes, T. Optical control of fish and zooplankton populations. *Limnol. Oceanogr.* **2004**, *49*, 233–238.

38. Walker, I. Amazonian Streams and Small Rivers. In *Limnology in Brazil*; Tundisi, J.G., Bicudo, C.E.M., Matsamura-Tundisi, T., Eds.; Brazilian Academy of Sciences: Rio de Janeiro, Brazil, 1995; pp. 167–194.

39. Lau, D.C.P.; Leung, K.M.Y.; Dudgeon, D. What does stable isotope analysis reveal about trophic relationships and the relative importance

of allochthonous and autochthonous resources in tropical streams? A synthetic study from Hong Kong. *Freshwater Biol.* **2009**, *54*, 127–141.

40. Brito, E.F.; Moulton, T.P.; de Souza, M.L.; Bunn, S.E. Stable isotope analysis indicates microalgae as the predominant food source of fauna in a coastal forest stream, south-east Brazil. *Austral Ecol.* **2006**, *31*, 623–633.

41. Newcombe, C.P.; MacDonald, D.D. Effects of suspended sediments on aquatic ecosystems. *North Amer. J. Fish. Manage.* **1991**, *11*, 72–82.

42. Waters, T.F. *Sediments in Streams—Sources, Biological Effects, and Control*; American Fisheries Society: Bethesda, MD, USA, 1995; p. 250.

43. Kleeberg, A.; Köhler, J.A.N.; Sukhodolova, T.; Sukhodolov, A. Effects of aquatic macrophytes on organic matter deposition, resuspension and phosphorus entrainment in a lowland river. *Freshwater Biol.* **2010**, *55*, 326–345.

44. Brookes, A. Response of aquatic vegetation to sedimentation downstream from river channelisation works in England and Wales. *Biol. Conserv.* **1986**, *38*, 351–367.

45. Wantzen, K.M.; Junk, W.J. Aquatic-terrestrial linkages from streams to rivers: Biotic hot spots and hot moments. *Arch. Hydrobiol. Suppl.* **2006**, *158*, 595–611.

46. Horeau, V.; Cerdan, P.; Champeau, A.; Richard, S. Importance of aquatic invertebrates in the diet of rapids-dwelling fish in the Sinnamary River, French Guiana. *J. Trop. Ecol.* **1998**, *14*, 851–864.

47. Odinetz Collart, O.; Jégu, M.; Thatcher, V.; Tavares, A.S. Les prairies aquatiques de l'Amazonie bresilienne. *RSTOM Actualités* **1996**, *49*, 8–14.

48. Cope, W.G.; Bringolf, R.B.; Buchwalter, D.B.; Newton, T.J.; Ingersoll, C.G.; Wang, N.; Augspurger, T.; Dwyer, F.J.; Barnhart, M.C.; Neves, R.J.; *et al.* Differential exposure, duration, and sensitivity of unionoidean bivalve life stages to environmental contaminants. *J. North Amer. Benthological Soc.* **2008**, *27*, 451–462.

49. Wantzen, K.M.; Pinto-Silva, V. Uso de substratos artificiais para macroinvertebrados bentônicos para a avaliação do impacto de

assoreamento em nascentes dos tributrios do Pantanal do Mato Grosso, Brasil. *Revista Brasileira de Recursos Hídricos* **2006**, *11*, 99–107.

50. Bash, J. *Effects of Turbidity and Suspended Solids on Salmonids*; Center for Streamside Studies, University of Washington: Seattle, WA, USA, 2001; p. 74.

51. Lloyd, D.S. Turbidity as a water quality standard for salmonid habitats in Alska. *North Amer. J. Fish. Manage.* **1987**, *7*, 34–45.

52. Bruton, M.N. The effects of suspensoids on fish. *Hydrobiologia* **1985**, *125*, 221–242.

53. Gende, S.M.; Edwards, R.T.; Willson, M.F.; Wipfli, M.S. Pacific salmon in aquatic and terrestrial ecosystems. *BioScience***2002**, *52*, 917–928.

54. Carolsfeld, J.; Harvey, C.; Ross, C.; Baer, A. *Migratory Fishes of South America: Biology, Fisheries and Conservation*; International Development Centre & the World Bank: Ottawa, Ontario, Canada, 2003; p. 372.

55. Cederholm, C.J.; Salo, E.O. *The Effects of Logging Road Landslide Siltation on the Salmon and Trout Spawning Gravels of Stequaleho Creek and Clearwater River Basin, Jefferson County, Washington, 1972–1978*; Fisheries Research Institute, University of Washington: Seattle, WA, USA, 1979; p. 90.

56. Van der Sluijs, I.; Gray, S.M.; Amorim, M.C.P.; Candolin, U.; Hendry, A.P.; Krahe, R.; Maan, M.F.; Utne-Palm, A.C.; Wagner, H.J.; Wong, N.B.M. Communication in troubled waters: Responses of fish communication systems to changing environments. *Evol. Ecol.* **2011**, *25*, 623–640.

57. Keenleyside, M.H.A. Some aspects of schooling in fish. *Behaviour* **1955**, *8*, 183–249.

58. Keenleyside, M.H.A.; Bietz, B.F. The reproductive behavior of *Aequidens vittatus* (Pisces, Cichlidae) in Surinam, South America. *Environ. Biol. Fish.* **1981**, *6*, 87–94.

59. Seehausen, O.; Van Alphen, J.J.M.; Witte, F. Cichlid fish diversity threatened by eutrophication that curbs sexual selection. *Science* **1997**, *277*, 1808–1811.

60. Dias, A.M.; Tejerina-Garro, F.L. Changes in the structure of fish assemblages in streams along an undisturbed-impacted gradient, upper Parana River basin, Central Brazil. *Neotrop. Ichthyol.* **2010**, *8*, 587–598.

61. Mol, J.H.; You, K.W.T.; Vrede, I.; Flynn, A.; Ouboter, P.; van der Lugt, F. Fishes of Lely and Nassau Mountains, Suriname. In *a Rapid Biological Assessment of the Lely and Nassau Plateaus, Suriname (with Additional Information on the Brownsberg Plateau)*; Alonso, L.E., Mol, J.H., Eds.; Conservation International: Arlington, TX, USA, 2007; pp. 107–118.

62. Lujan, N.K.; Roach, K.A.; Jacobsen, D.; Winemiller, K.O.; Vargas, V.M.; Ching, V.R.; Maestre, J.A. Aquatic community structure across an Andes-to-Amazon fluvial gradient. *J. Biogeogr.* **2013**.

63. Moyle, P.B.; Leidy, R.A. Loss of Biodiversity in Aquatic Ecosystems: Evidence from Fish Faunas. In *Conservation Biology*; Fiedler, P.L., Jain, S.K., Eds.; Chapman & Hall: New York, NY, USA, 1992; pp. 127–169.

64. Burcher, C.L.; McTammany, M.E.; Benfield, E.F.; Helfman, G.S. Fish assemblage responses to forest cover. *Environ. Manage.* **2008**, *41*, 336–346.

65. Casatti, L.; Ferreira, C.P.; Carvalho, F.R. Grass-Dominated stream sides exhibit low fish diversity and dominance by guppies: An assessment of two tropical pasture river basins. *Hydrobiologia* **2009**, *632*, 273–283.

66. Wantzen, K.M.; Nunes da Cunha, C.; Siqueira, A.J.B. Cerrado Stream Valleys and their Vegetation: Structure, Impacts by Erosion and Recuperation Strategies. In *the Pantanal: Ecology, Biodiversity and Sustainable Management of a Large Neotropical Seasonal Wetland*; Junk, W.J., da Silva, C.J., Nunes da Cunha, C., Wantzen, K.M., Eds.; Pensoft: Moscow and Sofia, Russia, 2011; pp. 143–165.

67. Valentin, C.; Poesem, J.; Li, Y. Gully erosion: Impacts, factors and control. *Catena* **2005**, *63*, 132–153.

68. Hammond, D.S.; Rosales, J.; Ouboter, P.E. *Managing the Freshwater Impacts of Surface Mining in Latin America*; IDB Technical Note 519; Inter-American Development Bank: Washington, DC, USA, 2013; p. 36.

69. Wang, J.J.; Lu, X.X.; Liew, S.C.; Zhou, Y. Retrieval of suspended sediment concentrations in large turbid rivers using Landsat ETM+: An example

from the Yangtze River, China. *Earth Surf. Process. Landf.* **2009**, *34*, 1082–1092.

70. CARBIOCIAL (Carbon Sequestration, Biodiversity and Social Structures in Southern Amazonia: Models and Implementation of Carbon-Optimized Land Management Strategies) Project. Available online: http://www.carbiocial.de/ (accessed on 24 September 2013

# Chapter 2

# Assessing Farmers' Perception of Soil Erosion Risk in Northern Jordan

**Yahya Farhan, Dalal Zregat, Ali Anbar**

Department of Geography, University of Jordan, Amman, Jordan

## ABSTRACT

Socioeconomic factors and farmer's perception of soil erosion and conservation were examined with special reference to Wadi Kufranja catchment, northern Jordan. Field data were collected through a household field survey, and soil erosion loss was calculated and mapped using the Revised Universal Soil Loss Equation (RUSLE), within a GIS/RS environment. In-situ field measurements of soil erosion were also conducted to assess splash, sheet and runoff soil erosion. The estimated potential average annual soil loss is 10 ton·ha$^{-1}$ year for the watershed. 42.1% (5317. 23 ha) of the watershed area was estimated to have moderate soil loss (5 - 25 ton·ha$^{-1}$·years $^{-1}$). Soil erosion risk is severe to extreme over 31.2% (3940. 56 ha) of the catchment, whereas the calculated soil loss is 25 - 50 and >50 ton·ha$^{-1}$·year$^{-1}$. The measured sheet and splash soil erosion in W. Kufranja was 10 ton·ha$^{-1}$·year$^{-1}$ from tillage land, and 3 ton·ha$^{-1}$·year$^{-1}$ from the fallow land, with an average ranges from 8 to 10 ton·ha$^{-1}$·year$^{-1}$. Similarly, the maximum measured soil erosion on the eastern margin of W. Kufranja was 12.7 ton·ha$^{-1}$·year$^{-1}$, while the minimum soil erosion was 2.9 ton·ha$^{-1}$·year$^{-1}$. The collected household socioeconomic/conservation data have been subjected to multivariate statistical analysis. Through factor analysis, the twenty one variables were reduced into four significant factors which account for 69.7% of the variation in the original variable. Stepwise multiple regression analysis revealed that the total variance explained by three independent variables was 0.585 ($R = 0.765$, $R^2 = 0.585$). Out of the total variance, forest clearance explained 34.7%, fallow land 7.7%, and land use/land cover 16.1% respectively. The F-value for forest clearance, fallow land, and land use/land cover are significant at 0.1% level. Most of the farmers aware that poor land management, deforestation, overgrazing, traditional cultivation (cultivation up-and-down the slope, and mono-cropping), and population pressure, are the major direct and indirect causes of soil erosion. By contrast, vegetative measures (i.e., afforestation and tree planting), adoption of structural soil and water conservation measures (terraced farming, check dams and gully control), and crop system management were recommended to control soil erosion.

## KEYWORDS

Soil Erosion Perception, GIS, Terrain Units, Soil Conservation, Factor Analysis, Stepwise Multiple Regression

## 1. INTRODUCTION

Soil degradation by accelerated erosion is a serious environmental problem in the highlands region of Jordan. Erosion of topsoil leads to declining soil quality and productivity, thus restricting the area of potential future agriculture. Rapid population growth, the associated population movements into rural areas, and land use practices and land cover changes since the 1950's, have led to soil degradation. Eroded soil materials from the denudational slopes are deposited over wadi floors, agricultural lands, irrigation canals, even on roads, and more seriously in reservoirs constructed across the wadis draining to the rift. Several studies were carried out to estimate soil erosion in Jordan on local, regional, and national scales [1] -[3]. In situ field measurements of runoff soil erosion and splash erosion were carried out (1985/1986 and 2003/2004) in the upper and lower Wadi Kufranja and Jerash area close to the wadi [4] [5]. However, the main causes of land degradation in Jordan have been outlined at regional [6], and national levels [7]. It can be concluded that the principal causes of land degradation and desertification in Jordan are: soil erosion by water and wind, land use changes, poor vegetal cover, exploitation of underground water and soil salinity, over grazing, and rapid urbanization. Thus, loss of soil fertility and productivity are the main consequences of land degradation in Jordan [8].

Recently, the revised Universal Soil Loss Equation (RUSLE) model has been employed in conjunction with RS and GIS technology to estimate the annual soil loss in different areas in central and northern Jordan [9] [10] and southern Jordan, [11]. Moreover, Alkharabsheh et al., [12] assessed the impact of land cover changes on soil erosion hazard in the southern part of the Yarmouk catchment using remote sensing and GIS. They concluded that the differences in erosion risk between 1992 and 2009 were considerable, and mainly attributed to changes in land cover/land use which influenced soil erosion rates significantly in northern Jordan. The influence of physical environmental factors on soil erosion loss was investigated recently by Farhan et al., [13], and it was reported that soil erosion was sensitive to specific physical factors such as slope, elevation, aspect, and terrain units. In the present investigation, socioeconomic determinants and farmers' perceptions of soil erosion and soil conservation measures were explored. Perception in this context is recognized as the process by which individuals and communities explain and organize their impressions to create a meaningful experience of the real world. Simply, it refers to the way communities try to understand the environment, and based on their experience with land resources, how they exploit this in making their living [14] [15]. The data collected through the household survey, and other data on estimated/measured soil erosion [4] [5] [10] were utilized to assess the status and people's perception of soil erosion, and to identify socioeconomic determinants of soil erosion, through factor analysis and stepwise multiple regression.

## 2. STUDY AREA

Wadi Kufranja catchment constitutes the present study area. It is located in the northern highlands of Jordan, and lies between 32°14' to 32°22' north latitudes and 35°21' to 35°47' east longitudes (Figure 1). The watershed area is 126.3 km$^2$ (12,630 ha), with elevations 1137 m a.s.l (above sea-level) to −329 m b.s.l (below sea-level) over a short distance of only 23 km. The upper wadi consists of maturely dissected terrain, with broad valley forms and smooth interfluves. In the middle and lower parts, rejuvenation resulted in a narrow, incised, steep-sided gorge. The longitudinal profiles of northern Jordan wadis generally exhibit four major irregularities. It is believed that these irregularities represent rejuvenation stages that took place during the Oligocene-Miocecne and Pleistocene. The flat summits around Ajlune, Anjara and Kufranja towns represent the remnants of Tertiary and/or Pleistocene erosion surfaces [16] -[18]. Slope category of 0° - 5° is restricted to the wadi alluvial fan (the Ghor), west of Krayma and scattered summit surface east of Krayma, meanwhile slope categories of 5° - 10° and 10° - 15° dominate in the scattered narrow strips which constitute the remnants of erosion surfaces, the structural benches and the upper convex slopes. The steep slope categories of 15° - 20°, 20° -30° and >30° characterized most lower wadi-side slopes in the upper, middle, and lower parts of the watershed. Large and small scars and hummocky topography indicate past landside events, probably of Pliocene and Quaternary age (<5 Ma) [16] [19]. The geology is dominated by Upper Cretaceous marly clay and marly limestone of the Ajlun series, which closely influences the basin soils. Red Mediterranean shallow soils ("terra rossa") cover the largest area in the watershed, while other types comprise brown limestone soils of the limestone outcrop, Rendzinas, alluvial (wadi) soils, variable types on slopes, and soils of the alluvial fan at the bottom of Wadi Kufranja, west of Krayma town. About 10% of the watershed is bare rock, and truncation of upper soil horizons is widespread; thus fully developed soil profiles are rare. Erosion exposes more loosely structured subsoil, which accelerates further erosion [20]. Climate is of the "dry Mediterranean" type in the upper catchment and "semi-arid and arid" types in the lower parts. Mean annual rainfall ranges from 630.5 mm (Ajlune station) to 267.8 mm (Wadi Kufranja station, east of Krayma) close to the Ghor. 95% of the precipitation falls from November to March (70% in Dec-Feb).Winter monthly temperatures of 3°C - 5°C are recorded in higher parts of the watershed; summer months average 22° - 25°C. In Krayma (in the Ghor, the Jordan Rift-floor) the average annual temperature is 24°C, with summer months reaching 40+°C. Frost-days number 5 - 15 per year [17] [21].

Land-cover types vary from natural vegetation (forests) mixed with crop-land. Four scattered associations of forests are distinguished throughout the watershed [22]: broad-leaved forest of Kermes oak, broad-leaved forest of decideous oak (or) coniferous forest of Aleppo pine, and mixed forest of oak and wild olive. The land use / cover of the watershed comprises 7.52% residential and commercial areas, 18.53% forest, 51.23% mixed agricultural, 13.30% open rangeland and 9.42% bare soil [23]. The forests and other vegetation types in northern Jordan have suffered severe anthropogenic interventions. Overgrazing of rangeland (mainly by goats), and the collection of fuel-wood and charcoal-wood are the main causes underlying forest degradation. The effect of cultivation patterns and land cover changes is that the soil surface is bare during the moist winter months. Low rainfall interception by vegetation allows destructive splash erosion [20]. At present, shallow landslides, occasionally deep-seated landslides, soil slumping, and minor mudflows are repetitive phenomena following intense rainfall. Previous investigations [19] [24] -[29], indicate that such movements are triggered primarily by the combined influence of hill-slope disturbance by intensive human activities, soft rock formation, poor land management, and recurrent heavy/intense rainstorms. Such processes increase soil erosion rates.

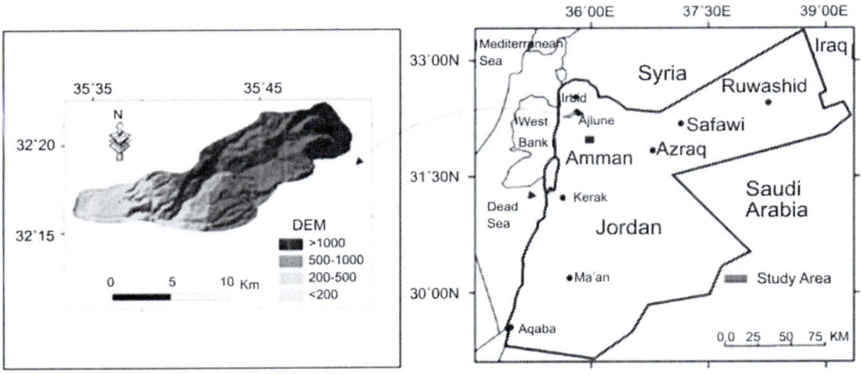


**Figure 1**. The study area.

## 3. MATERIALS AND METHODS

Data were obtained from a variety of sources including remote sensing, DEM, and GIS for estimating soil erosion loss; field measurements of soil erosion, and from a household questionnaire survey. The basic methods/ tools employed to assess soil erosion are illustrated (Figure 2) as follows:

### 3.1. Soil Erosion Estimation

Soil erosion in northern Jordan was conducted over the past two decades using different soil erosion estimation techniques such as: in-situ field measurements, and soil erosion estimation using geospatial technology (remote sensing and GIS).

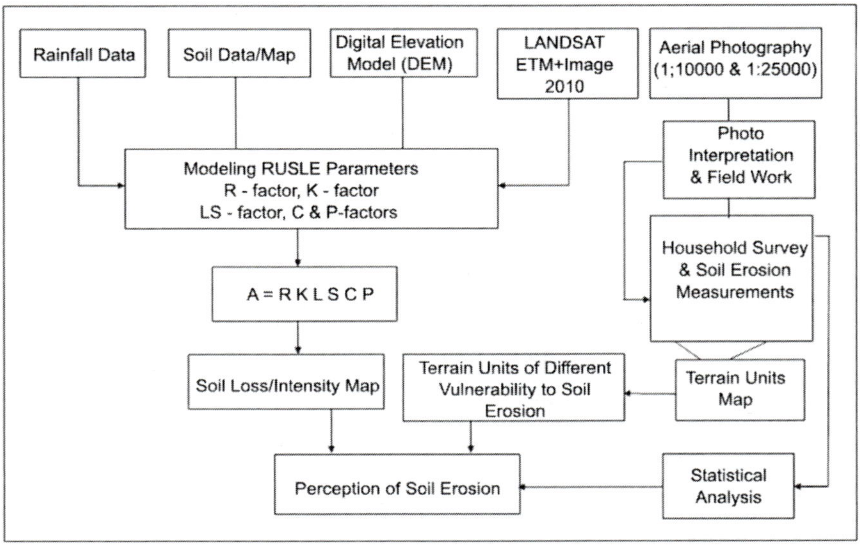


**Figure 2**. Methodology of the study.

## 3.1.1. Estimation of Soil Erosion Loss Using the RUSLE Approach

Several different empirical and physical process based models have been developed since 1930s to predict soil loss [30] at local, watershed, and regional scales. Recently the Revised Universal Soil Loss Equation (RUSLE) [31] -[34] is considered the most effective model employed to estimate soil loss worldwide. It uses elaborate methods to compute soil erosion factors, predicting annual soil loss by rain-splash and surface runoff over all types of land-use and cover-type. The protocol is considered essential for planning soil conservation [35]. Five parameters are manipulated in the model, in raster format: rainfall erosivity, soil erodibility, slope length and steepness, cover-type and management, and support practice. The equation is written as:

$$A = R \times K \times LS \times C \times P$$

where,

A: the computed soil loss per unit area [ton·ha$^{-1}$·year$^{-1}$], R: runoff erosivity factor (rainfall and snowmelt) in [MJ mm ha$^{-1}$·hr $^{-1}$·year$^{-1}$], K: soil erodibility factor (soil loss per erosion index unit for a specified soil measured on standard plot, 22.1m long, with uniform 9% (5.16°) slope, in continuous tilled fallow) [ton·ha·hr·ha$^{-1}$ MJ$^{-1}$·mm$^{-1}$], L: slope length factor (ratio of soil loss from the field slope length to soil loss from standard 22.1m slope under identical conditions) (dimensionless), S: slope steepness factor (ratio of soil loss from the field slope to that from the standard slope under identical conditions) (dimensionless), C: cover-management factor (ratio of soil loss from a specified area with specified cover and management to that from the same area in tilled continuous fallow) (dimensionless), and P: the support practice factor (ratio of soil loss with a support practice-con- tour tillage, strip-cropping, terracing-to soil loss with row tillage parallel to the slope (dimensionless).

The RUSLE model was applied with Arc GIS 10.1 and ERDAS Imagine 8.5, along with LANDSAT ETM+ 2010 to estimate soil erosion loss [10] [11]. Aerial photographs of scale 1:10,000 (1978), and 1:30 000 (1992) were utilized to compile the terrain units map through photo-interpretation methods and field survey [36]. Land use/cover information for the watershed was revised and updated using Google Earth Pro (2011). Rainfall data for calculation of rainfall erosivity (R) were obtained from the Ministry of Water and Irrigation, and the soil data were acquired from 1995 national soil survey maps/reports [37]. NDVI values generated and mapped from a LANDSAT image were used to determine the C factor, and to verify land use/cover information. The GIS Tool Box was employed to compute the parameters of the RUSLE model and to compile soil erosion loss map.

### 3.1.2. Field Measurements of Soil Erosion

Two plots for field measurements of sheet, runoff (soil traps) and splash soil erosion [38] [39] were established in W. Kufranja [4], one plot installed on tillage land, and the other assigned on fallow cultivated land (Figure 3). Field measurements were conducted during the winter of 1985/1986. Beni Taha [5] also, carried out field measurements to estimate runoff and splash soil erosion on the eastern margins of W. Kufranja to the west of Jerash city. Two of the four plots were installed close to the eastern margin of W. Kufranja watershed. The measurements were carried out during the winter of 2002/2003. In both case studies, the physical factors affecting soil erosion were evaluated using stepwise regression analysis. The selected sites for soil erosion measurements varied in terms of morphological characteristics (slope form (convex, concave, or straight), gradient (degrees); aspect, length of slope (m), location on slope (upper, middle, or lower), vegetation and land use, conservation measures, fallow or tillage (contour ploughing, or up-and-down slope ploughing), stone abundance%, and soil texture.

(a)

(b)

(c)

**Figure 3**. Soil erosion equipments installed to measure sheet, rain splash, and runoff erosion.

### 3.1.3. Terrain Unit Mapping

The terrain mapping unit (TMU) (or homogenous domain) approach was elaborated by Meijerink [36] as a method for data acquisition and storage. It makes use of the interactions between geology, geomorphology, hydrology, vegetation and soils. For the purpose of the present investigation definitions of terrain units proposed by Meijerink [36] were adopted as given below:

A TMU represents a natural division (or) portion of land surface containing a set of ground conditions which differ from the adjoining units, and it's boundaries can be easily demarcated [40] [41]. The unit can also be recognized on aerial photographs, and verified by field check. Terrain units are normally differentiated on the basis of photographic properties and landform attributes as they appear and can be recognized in the stereo model and the field. A TMU is also identified as a unit where a group of interrelationships among landform, lithology and soil are dominant. Moreover, a TMU differs from another adjacent unit because either the landforms are evidently different, or the phenomena associated with the landform differ. Thus, and in terms of GIS, a TMU may be described as the geographic location (polygons) of entities which are related to a unique set of attributes. Terrain unit mapping was accomplished based on the interpretation of aerial photographs associated with a field check. The landform type, hydrology, vegetation and soil conditions of the unit were considered essential in the process of mapping these units. A terrain unit is thus necessary to establish a data-base of the environmental data needed to describe a given process or phenomenon, and to build up a predictive or explanatory model of the spatial or temporal distribution of such a phenomenon. The procedures of terrain units mapping using remote sensing techniques are discussed in details by van Zuidam [42] [43], Verstappen [44], Speight [40] and Meijerink [36].

### 3.1.4. Famers' Perceptions and Socioeconomic Determinants of Soil Erosion: Household Survey

The study watershed consists of 2295 farms distributed over 18 villages and towns [45]. Six villages accommodate between 5 and 10 farms, therefore the farm sample was drawn from the remaining 12 villages. The sample size was determined according to the following Yamane's formula [46]:

$$n = \frac{N}{\left(1 + N \times \mathrm{e}^2\right)} \tag{1}$$

where:

n is sample size, N is total no. of households and e is the significance level (in our case the significance level is 7%). The resultant sample size was 187 farms which represent the whole watershed. A structured questionnaire was designed and distributed to collect information on socioeconomic conditions of farmers, soil erosion status, soil and water conservation, and the perception of soil erosion extent, causes and impacts, and the awareness of soil conservation measures and practices. The information also covers the household characteristics, land ownership, income and expenditure, and the awareness of governmental soil erosion projects including cost. The collected information is considered an essential factor when making decisions on soil and water conservation projects and practices [47]. Two categories of information were collected by means of the questionnaire survey (Table 1): the socioeconomic variables/ determinants of soil erosion, farmers' perceptions of soil erosion, soil conservation types, cost, and effectiveness of installed measures across the watershed. However, appropriate questions were included in the questionnaire and aimed to reveal the direct and indirect causes of soil erosion as perceived by the farmers. Also, a final question asked the farmers to record their suggestions regarding the appropriate conservation measures needed to minimize soil erosion in W. Kufranja.

Factor analysis was employed to condense the original variation, measured in terms of a large number of variables (21 variables), into variation in terms of a few factors, each of which contributes a known amount to the total variation It is performed on the standardized variables using correlation matrix to dismiss the effect of different measurement units on the determination of factor loadings. Factor loadings represent a simple correlation between properties of each factor. Eigenvalues illustrate the amount of variance explained by each factor. Factors with eigenvalues > 1 explained more the total variation in the data than farmers' socioeconomic characteristics and perception of soil erosion, while factors with eigenvalues <1 explained less the total variation than the individual farmer. Thus, only factors with eigenvalues >1 were considered for interpretation of results [48] [49]. Table 1 illustrates the 21 variables with their characteristics, and employed in factor and stepwise regression analysis. The soil erosion rate for each farm was derived from the soil erosion loss map, and considered the dependent variable.

Some important socioeconomic variables were eliminated in the present analysis. All households in W. Kufranja for example, are owned only by males. Therefore, the gender variable does not contribute to variation in perception. All farmers also inherited their farms through successive generations, thus, household size is generally small (1.5 - 2.5 ha only) due to continuous fragmentation of the agricultural lands. This process is expected to affect long-term investment in soil conservation activity, and encourage soil erosion. Hence, the law of agricultural land parceling and the resultant lands fragmentation has been excluded from the analysis, since most households are homogeneous in terms of size, and cultivation practices.

# 4. RESULTS AND DISCUSSION

## 4.1. Soil Erosion Rates

The data layers (maps) extracted for K, LS, R, C, and P factors of the RUSLE model were integrated within the raster calculator option of the Arc GIS spatial analyst in order to quantify, evaluate, and generate the soil erosion loss map for Wadi Kufranja [10]. Such an assessment is essential to examine the farmers' perception of soil erosion loss, and soil conservation measures, and to evaluate the socioeconomic factors of soil erosion for future agricultural and soil conservation planning. The magnitude and spatial distribution of crop management factor C show values between 0.01 and 0.2. The highest (poor land cover management) almost coincide with the lowest NDVI values, (0.22 - 0.05), since the forest protects soils against erosion, while the open rangeland exposed to plowing has a high C-value (0.35). Similarly, the mixed rain-fed areas have a C-value of (0.2). The model showed logical results after applying the assumed C values for each land-cover class, with a trend of increasing erosion with low vegetation cover. Moreover, it was possible to reveal three classes of P factor by considering land use/cover-type and support factors. P factor ranges from 0.19 to 1.0, the higher values in areas east of Krayma with no conservation practices (forest/natural vegetation), and other major settlements in the catchment. By contrast, maximum P values correspond to crop-land with contour, and terrace tillage in the upper catchment. P values decrease towards the upper catchment, where in flat land units slope length decreases. This explained lower P values in the irrigated lands compared to open rangeland, since irrigated farms occupy mainly flat/ undulating lands. Average annual soil loss of 10 $ton \cdot ha^{-1} \cdot year^{-1}$ was estimated for the whole catchment, and the final soil loss map indicates a minimum P value of 0.0 to a maximum of 1865 $ton \cdot ha^{-1} \cdot year^{-1}$ [10]. The Wadi Kufranja watershed was classified into five soil erosion loss classes, and the potential soil erosion loss (Table 2) increases from the upper to the lower reaches of the catchment. Soil erosion is very severe east of the Krayma area and accounts for 31.2% of the total watershed soil loss. The upper part of the watershed is well forested with some mixed agricultural cover, hence it is of low erosion loss. The distribution of soil erosion loss classes (Figure 4) show that 26.7% of the watershed has minimal soil loss, 36.5% is low, 5.6% and 7.9% is moderate and severe respectively, while 23.3% of the watershed is exposed to extreme soil erosion.

**Table 1**. Socioeconomic and soil conservation variables.

| Socioeconomic/soil conservation variables | Measurement level | Value |
|---|---|---|
| Dependent variable/Soil erosion rate | $ton \cdot ha^{-1} \cdot year^{-1}$ | Continuous |
| Covariates: | | |
| X1. Age (years) | Number | Discrete |
| X2. Educational level | ≥Grade 6 | Discrete |
| X3. Distance between house and the farm | Km | Continuous |
| X4. Size of agricultural holding | Dunum | Discrete |
| X5. Total of farm capital | Assets/JD | Continuous |
| X6. Farm annual economic/revenue | Total household income/JD | Continuous |
| X7. Farm annual expenditure | Farm expenditure/JD | Continuous |
| X8. Farm labor | Number | Discrete |
| X9. Farm accessibility | Easy/difficult | Dummy |
| X10. Dependency ratio | Number | Discrete |
| X11. Adopted method(s) of soil conservation | 0 (or) 1 | Dummy |
| X12. Indigenous knowledge/practical experience in soil conservation | 0 (no), 0.3 (low), 0.6 (moderate), 1 (high) | Continuous |

| | | |
|---|---|---|
| X13. Proper training in soil conservation | Frequency/year (or)years | Discrete |
| X14. Awareness of soil conservation cost | 1 (yes), 0 (no) | Dummy |
| X15. Beneficial from governmental conservation projects | 1 (yes), 0 (no) | Dummy |
| X16. Does government help farmers in their conservation efforts | 1 (yes), 0 (no) | Dummy |
| X17. Are you willing to participate in conservation measures cost | 1 (yes), 0 (no) | Dummy |
| X18. Does forest clearance accelerate soil erosion | 1 (yes), 0 (no) | Dummy |
| X19. Does land cover and land use changes accelerate soil erosion | 1 (yes), 0 (no) | Dummy |
| X20. Does fallow land accelerate soil erosion | 1 (yes), 0 (no) | Dummy |
| X21. Does climate change accelerate soil erosion | 1 (yes), 0 (no) | Dummy |

The results of the present investigation in the Wadi Kufranja watershed are comparable with similar studies carried out in north Jordan [10] [11] [13] [50], where similar terrain, climatic conditions, and historic land mismanagement prevail. Moreover, the estimated soil erosion levels are consistent with those results reported from other Eastern Mediterranean watersheds using the RUSLE approach, where relatively similar environmental conditions exist [51] -[54]. The present results emphasize the necessity for well executed research on soil erosion and improved conservation methods. The measured sheet and splash soil erosion in W. Kufranja was 10 ton·ha$^{-1}$ year$^{-1}$ from tillage cultivated lands, and 3 ton·ha$^{-1}$·year$^{-1}$ from the fallow land, with average ranges from 8 - 10 ton·ha$^{-1}$·year$^{-1}$ [4]. Similarly, the measured maximum soil erosion on the eastern margin of W. Kufranja was 12.7 ton·ha$^{-1}$·year$^{-1}$, while the minimum soil erosion measured was 2.9 ton·ha$^{-1}$·year$^{-1}$. It is concluded that vegetation and land use, rainfall intensity during the storm event (mm/hr$^{-1}$), the amount of rainfall for the storm event (mm), conservation measures, and slope form, are the most variables influencing soil erosion [5].

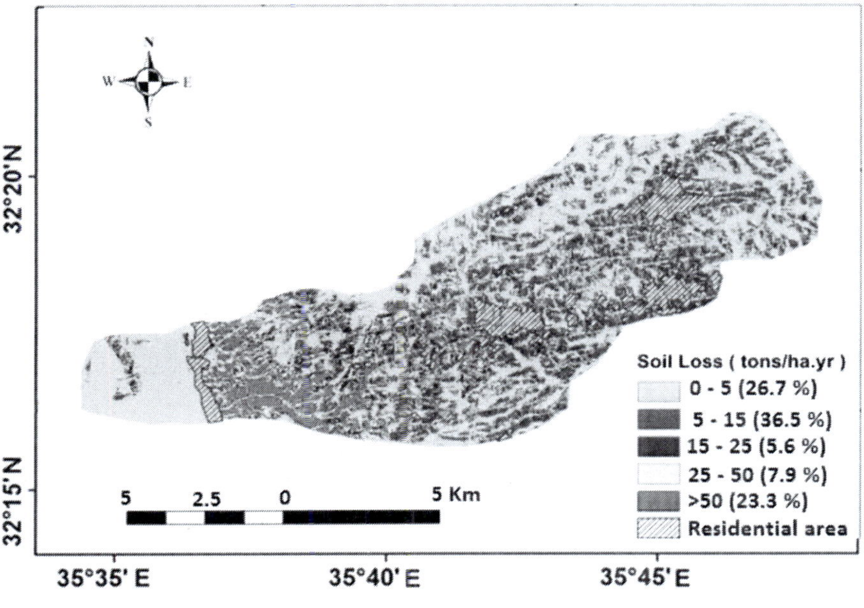


**Figure 4**. Soil erosion loss categories

**Table 2**. Area and proportion of each soil loss classes.

| Erosion loss class | Numeric range ($ton \cdot ha^{-1} \cdot year^{-1}$) | Percentage (%) | Area (ha) |
|---|---|---|---|
| Minimal | 0 - 5 | 26.7 | 3372.21 |
| Low | 5 - 15 | 36.5 | 4609.95 |
| Moderate | 15 - 25 | 5.6 | 707.28 |
| Severe | 25 - 50 | 7.9 | 997.77 |
| Extreme | >50 | 23.3 | 2942.79 |

Source: Farhan *et al.*, (2013).

The measured and calculated soil erosion per year from the different plots in Jerash area are comparable to those estimated by adopting the RUSLE model [10], and the suspended sediment yield calculated using Fournier's model [4]. Such comparison indicates the reliability of the RUSLE model in estimating soil erosion rates in northern Jordan. However, all previous studies concluded that the RUSLE model, combined with RS and GIS technologies, and field measurements of soil erosion are efficient tools in providing the decision makers with essential information for water and soil conservation to maintain future agricultural sustainability. Yet to control soil erosion, conservation measures should be executed at the farm, hillslope or watershed scale (i.e. W. Kufranja catchment). Moreover, the implementation of conservation measures to specific terrain units subjected to severe soil erosion loss is of high priority. The terrain unit is considered an appropriate manageable land unit in both soil conservation and other applications [44]. Vegetation cover as well as slope, soil and land use/land cover is major impact factors in soil erosion. Combined with the generated map, this shows the spatial patterns of soil loss rates (Figure 4), all these variables can be employed to assess the priority areas for soil conservation [55]. Planning the future conservation interventions is another important target. Figure 5 illustrates the spatial distribution of 19 terrain units in W. Kufranja. It is obvious that 55.06% of soil erosion occurred on five terrain units of moderate and steep slopes [13]. The valley-side slopes exhibit the highest rate of soil erosion compared with flat terrain units. The alluvial terrain units experience the lowest rate (Table 3). It is noticeable that soil erosion is closely controlled by slope: the steeper the slope, the more severe the erosion. Terrain units of steep and long slopes characterized by low vegetation cover thus, exhibit much higher rates of erosion compared to flat/undulating terrain units [56] -[58].

The old landslide terrain unit is considered as degraded terrain, where it shows a low rate of soil loss generally, although relatively steep (10° - 15°, 15° - 20°, and 20° - 30° slope categories) and a high amount of rainfall are dominant. The soil erosion rate decreases here because the change in land use/cover from bare soil/rangeland to forest and shrubs stabilized the landslide areas. The flat summit terrain shows the lowest rate of soil erosion. Also, soil erosion is high on the remnants of erosion surfaces although flat to slightly sloping terrain are common (Table 3). This is explained by the dominance of "rainfed farming" practiced over these terrain units. It has been reported recently that "mixed rainfed" farming is one of the major reasons behind the high and very high soil erosion rates characterizing this terrain [12].

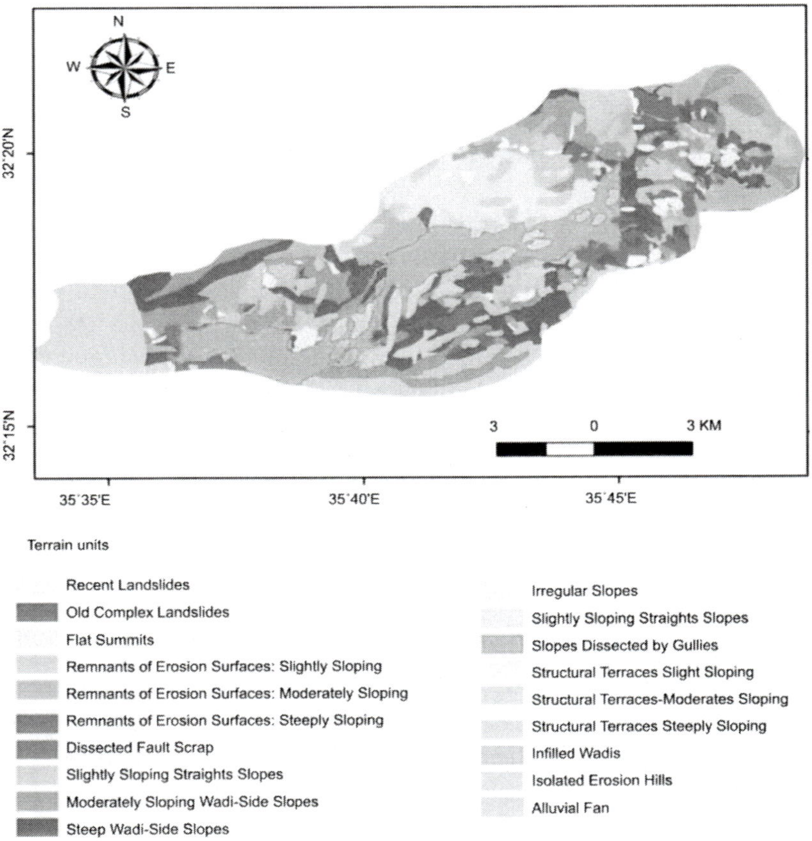


**Figure 5**. Terrain units of W. Kufranja.

## 4.2. Farmers' Perceptions of Soil Erosion

With reference to the perceptions of farm households participating in the present survey, it was possible to identify five major direct causes of soil erosion in W. Kufranja: poor land management (42.3% of the respondents), deforestation and grazing, land use abuse and traditional cropping system, urbanization and natural causes (Figure 6). As pointed out earlier, 36.5% of the watershed is exposed to low soil losses (<15 ton·ha$^{-1}$·year$^{-1}$), whereas 36.8% of the watershed has experienced moderate, severe, and extreme soil erosion (15 - >50 ton·ha$^{-1}$ year$^{-1}$. These very high soil erosion rates are attributed mainly to improper soil management, poor conservation practices, and traditional cultivation. Deforestation (for fuel and charcoal-wood), overgrazing, and destruction of vegetation cover were perceived by 32.6% of the respondents as causes of soil erosion, and land use/land cover change, tillage practices (up-and-down the slope)and traditional sole cropping by 13.6% of the respondents and, development of infrastructure, urbanization and natural erosion by 11.5%.

**Table 3**. Spatial variation of soil erosion with different terrain units in W. Kufranja.

| Terrain units | Soil Loss Categories Tones/ha/year$^{-1}$ (Area %) | | | | |
|---|---|---|---|---|---|
| | Minimal | Low | Moderate | Severe | Extreme |
| | 0 - 5 | 5 - 15 | 15 - 25 | 25 - 50 | >50 |
| Moderately Sloping Wadi-Side Slopes | 11.16 | 1 | 0.73 | 1.04 | 0.09 |
| Steep Wadi-Side Slopes | 9.6 | 0.58 | 0.88 | 1.24 | 0.07 |
| Slightly Sloping Wadi-Side Slopes | 5.16 | 0.26 | 0.21 | 0.07 | 0 |
| Remnants of Erosion Surfaces: Steeply Sloping | 5.07 | 0.17 | 0.23 | 0.23 | 0 |
| Remnants of Erosion Surfaces: Moderately Sloping | 9.04 | 0.17 | 0.27 | 0.2 | 0.07 |
| Alluvial Fan | 6.46 | 0 | 0.03 | 0 | 0 |
| Isolated Erosion Hills | 0.34 | 0.03 | 0.07 | 0.04 | 0 |
| Structural Terraces Slight Sloping | 0.83 | 0 | 0.02 | 0.03 | 0 |
| Structural Terraces-Moderates Sloping | 0.75 | 0.03 | 0.21 | 0 | 0 |
| Structural Terraces Steeply Sloping | 1.74 | 0.29 | 0.22 | 0 | 0 |
| Old Complex Landslides | 0.41 | 0 | 0.23 | 0.09 | 0 |
| Irregular Slopes | 0.4 | 0 | 0.16 | 0.03 | 0 |
| Slightly Sloping Straights Slopes | 6.47 | 0.41 | 0.35 | 0 | 0 |
| Slopes Dissected by Gullies | 9.41 | 0.77 | 1 | 2.26 | 1.13 |
| Flat Summits | 0.58 | 0.04 | 0.03 | 0 | 0 |
| Infilled Wadis | 0.64 | 0 | 0.07 | 0.06 | 0 |
| Recent Landslides | 0.84 | 0 | 0.03 | 0.02 | 0 |
| Remnants of Erosion Surfaces: Slightly Sloping | 9.9 | 0.57 | 0.58 | 0.09 | 0 |
| Dissected Fault Scrap | 0.13 | 0 | 0 | 0.14 | 0 |

Source: Farhan et al. [14].

Indirect causes of soil erosion are comparably quite crucial since, these severely affect soil erosion through direct causes. Population pressure, land tenure and continuous fragmentation of agricultural land, poverty, traditional farming, and education were perceived as indirect causes of soil erosion in the area under consideration (Figure 7). About 66.9% respondents perceived population pressure and land tenure as major indirect causes of soil loss, while poverty, traditional farming, and education were perceived as major indirect causes by 30.1% of the respondents. The respondents were asked to give their opinion about the effective soil measures for controlling soil erosion. Figure 8 illustrates the conservation measures suggested by farmers. 40.8% of respondents were convinced that afforestation and tree plantation are essential to reduce soil loss, while 38.5 of the farmers believed that structural solutions, such as the construction of stone terraces and check dams, the utilization of organic manure, controlling cropping intensity and fallow period are vital to conserve soil and water across different sloping terrain units of the watershed. Stone bunds are considered the most popular conservation measures adopted here, where historical soil erosion, intensive agriculture, and agricultural terraces are characteristic of the Levantine highlands (including Jordan) since the Iron Age [59] -[61]. 20.7% of respondents thought that efficient land management is needed to rehabilitate intensively exploited soil resources. The effect of conservation structures installed (mainly stone bunds-contour stone terraces, termed locally masateb, and check dams) during the 1970s of the last century [62] [63] caused a significant decline in soil erosion rates.

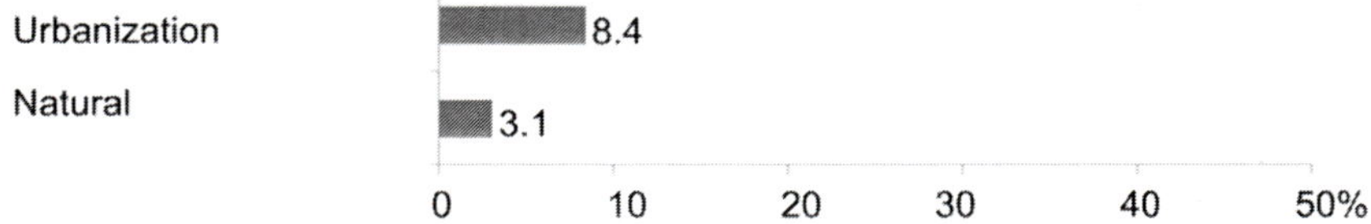


**Figure 6**. Direct causes of soil erosion.

Recent studies indicates that through applying soil conservation measures in central and northern Jordan, the annual erosion rate has decreased from 78 ton·ha$^{-1}$·year$^{-1}$, before installation of soil conservation structures, to 33 ton·ha$^{-1}$·year$^{-1}$, 20 years after installation of conservation structures [64]. Irrespective of installation of efficient conservation structures, improvement of cropping practice is essential to mitigate soil erosion.

Contour strip intercropping cultivation at a proper planting density (i.e. 350 plants $m^{-2}$) was found recently to be promising farming practice to reduce runoff and soil erosion [65]. Likewise, experimental results on soil erosion control and moisture conservation, showed that the presence of rock fragments on the soil surface were highly effective in reducing runoff and soil loss [66].

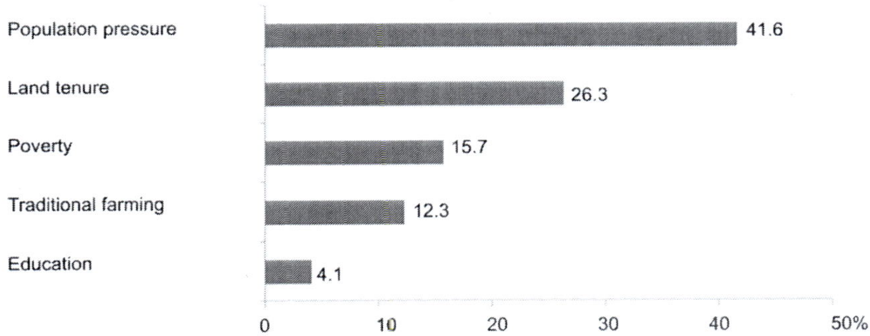


**Figure 7**. Indirect causes of soil erosion.

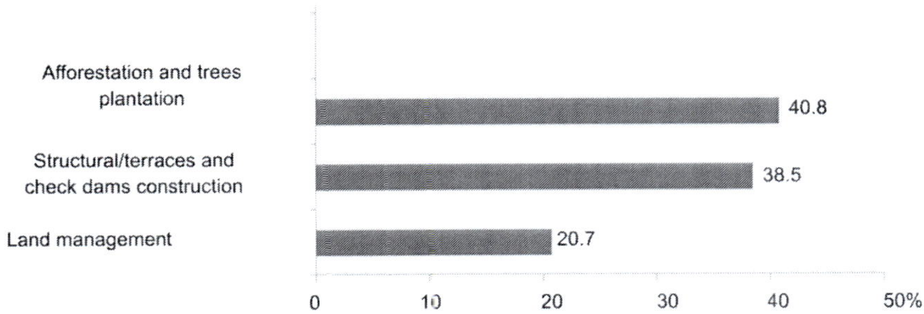


**Figure 8**. Suggested conservation measures to control soil erosion.

## 4.3. Socioeconomic Determinants of Soil Erosion

The relation between socioeconomic/soil conservation variables was examined using factor analysis and multiple stepwise regression. Factor analysis was employed to reduce the large number of variables, and to condense the original variation measured in terms of 21 variables, into variation in terms of a few factors, each of which contributes to a known amount to the total variation. Through factor analysis, the 21 variables were reduced into four significant factors which account for 69.7% of the variation in the original variables. Table 4 illustrates the cumulative contribution of the successive four rotated factors produced in the analysis.

The loadings of the rotated factors are shown in Table 5. Factor I exhibits relatively high positive loadings on five variables: age of farmers, distance, farm capital, farm revenue, and farm expenditure. By contrast, the variable farm labor is negatively loaded on factor I. Traditionally, the higher the number of family members, the more farm labor can be provided to help in farming and soil conservation activities. However, due to a continuous decrease in household size, most of the household owners are working at present in governmental jobs, leaving farm activities to workers hired on a temporary basis. Similarly, old farmers customarily used to con struct the conservation measures such as stone bunds, the most common structural measure in our region historically. By contrast, young farmers have abandoned this technique beginning in the 1970s. Factor I reflects the economic characteristics of the households. Factor II shows positive loading on forest clearance, dependency ratio, education, and adopted method(s) of conservation methods. Young and educated household owners are clearly aware of soil erosion hazards due to forest clearance and overgrazing. They are also familiar with different conservation methods installed in that area by the German Agricultural Technical Cooperation Team and the Jordanian Ministry of Agriculture Jordan four decades ago. Such conservation measures at that period of time were provided almost cost-free for each household. This may explain the farmers' unwillingness to invest in conservation measures, or to participate in paying for soil conservation costs. Factor II is considered an indicator of education and conservation knowledge, since the aforementioned four variables are correlated positively with factor II. The third factor accommodates positive loading on governmental help, conservation cost, and proper training in soil conservation. Therefore, it is considered as the government policy in soil conservation. Factor IV loaded positively on fallow land, and land use/land cover, hence, it is an indicator of land use/land cover changes. Factor IV is labeled as the land use/land cover factor. With the rotated factor solution illustrated above, the role of socioeconomic variables in household spatial variability was clearly explored, although, most farmers and farms are homogeneous in terms of several socioeconomic variables such as: farms ownership, gender, household size, land tenure, land parceling and the resultant land fragmentation.

**Table 4**. Relative importance of the first four factors.

| Factor Number | Eigenvalue | Percentage of variance | Cumulative percentage of total variance |
|---|---|---|---|
| I | 6.752 | 30.6 | 30.6 |
| II | 4.683 | 20.2 | 50.8 |
| III | 2.351 | 11.4 | 62.2 |
| IV | 1.761 | 7.5 | 69.7 |

**Table 5**. Basic factors and factor loading on socioeconomic/soil conservation variables.

| Factor no. | Factor lable | Factor loading |
|---|---|---|
| Factor I: | Farm economic characteristics | |
| Age | (1) | 0.461 |
| Distance | (3) | 0.646 |
| Farm capital | (5) | 0.610 |
| Farm revenue | (6) | 0.776 |
| Farm expenditure | (7) | 0.883 |
| Farm labor | (8) | −0.617 |
| Factor II: | Education and conservation knowledge | |
| Forest clearance | (18) | 0.562 |
| Dependency ratio | (10) | 0.551 |
| Education | (2) | 0.502 |
| Conservation methods | (11) | 0.488 |
| Factor III: | Government policy in soil conservation | |
| Government help | (16) | 0.636 |
| Conservation cost | (14) | −0.466 |
| Proper training | (13) | −0.454 |
| Factor IV: | Land use/land cover changes | |
| Fallow land | (20) | 0.571 |
| Land use, land cover | (19) | 0.439 |

Stepwise multiple regression analysis was employed to identify the predictor variables of soil erosion. (the dependent variable) among the socioeconomic/soil conservation variables (the independent variables). The model used is in the form given below:

$$Y_i = a + b_1X_1 + b_2X_2 + b_3X_3 + \cdots + b_jX_j \quad (2)$$

where:

$Y_i$ = soil erosion loss (ton·ha$^{-1}$·year$^{-1}$) estimated for each household parcel,

a = is a constant (the point where the line crosses the Y axis when X = 0,

b = regression coefficient,

$X_1, \cdots, X_j$ = the independent variables ($X_1$ - $X_{21}$).

The results of stepwise regression are illustrated in Table 6. From this table, the total variance explained by the three independent variables is 0.585 (R = 0.765, $R^2$ = 0.585). Out of the total variance, forest clearance explained 34.7%, fallow land 7.7%, and land use/land cover 16.1% respectively. The F value for forest clearance, fallow land, and land use/land cover are significant at 0.1% level. This indicates that 58 percent of variation in the dependent variables is explained by the recognized predictor variables (forest clearance, fallow land, and land use/land cover).

It is clear that anthropogenic factors such as: long and continuous human interference with land resources, deforestation and overgrazing in the past and present, land use changes and farming practices, poor conservation measures, and fragmentation of holdings are major causes of soil erosion [10] [20] [22]. Although household size is relatively large (6 - 8 members), most of the families do not provide farm labor as was the common practice in the past until the 1970s. Instead farm owners hire on the average three part-time workers to carry out necessary farm activities along with the restoration of old stone bunds and the checking of stone dams across the gullies on the farm. However, young farmers (<55 years old) are found to be more educated, and thus better aware of soil erosion problems, and more knowledgeable about soil conservation techniques and innovative farming practices, however, they are not willing to pay for or invest in soil conservation (construction of new facilities, or the renewal of older once practiced conservation measures) to reduce soil erosion. Irrespective of age, all farmers believe that the government should pay or invest in soil conservation installation. The household survey reveals that financial capital (average 45 250 JD), and farm income (average 3200 JD) seem adequate to facilitate the required credit needed to invest in soil conservation, or to improve existing structures. As farmers grow older and are unable to carry out agricultural activities, they tend to rent out their farms, in a situation where most of the younger generation also choose either higher education/professions, or join the army rather than practice farming. Moreover, farmers who cultivate land owned by others may be less likely to invest in soil conservation [47]. Tenants also lose part of their income as rent for the land, which acts as a serious financial obstacle for soil conservation, or the improvement of other aspects of the farm land. It can be concluded, that most farmers will abandon their farms in the future, although security of land tenure is very high in the W. Kufranja watershed.

# 5. CONCLUSIONS

The present results of the RUSLE model reveal the severity of soil erosion in Wadi Kufranja watershed. The mean soil loss estimated for the catchment was 10 ton·ha$^{-1}$·year$^{-1}$, with the five erosion risk classes, ranging from 0.0 to 1865 ton·ha$^{-1}$·year$^{-1}$. Areas of 53.1723 km$^2$ (5317.23 hectares) and 39.4056 km$^2$ (3940.56 hectares) were classed as suffering moderate or very severe soil erosion. Similarly, the measured sheet and splash soil erosion in W. Kufranja was 10 ton·ha$^{-1}$·year$^{-1}$ from tillage cultivated lands, and 3 ton·ha$^{-1}$·year$^{-1}$ from the fallow land, with an average ranging from 8 to 10 ton·ha$^{-1}$·year$^{-1}$. The measured maximum soil erosion on the eastern margin of W. Kufranja was 12.7 ton·ha$^{-1}$·year$^{-1}$, while the minimum soil erosion measured was 2.9 ton· ha$^{-1}$·year$^{-1}$. Vegetation and land use abuse, traditional cultivation, lack of conservation measures, and rainfall intensity for storm events (mm/hr$^{-1}$), are considered the most significant factors influencing soil erosion. Subsequently, several terrain units in the middle and lower reaches of the catchment must be prioritized for conservation, where high soil erosion rates predominate. Here, the combined effect of K, LS, and C factors, also accounts for high soil erosion loss across the study area. The household survey recognized poor land management, deforestation and overgrazing, land use abuse and traditional cropping system, urbanization and natural causes, as the major direct causes of soil erosion. Whereas, population pressure, land tenure and continuous fragmentation of agricultural land, poverty, traditional farming, and education were perceived as the major indirect causes of soil erosion.

**Table 6**. Regression of the estimated soil erosion on the independent variables.

| Variables | R | $R^2$ | Increase in $R^2$ % | F value for variable | Sig. % |
|---|---|---|---|---|---|
| Forest clearance | 0.589 | 0.347 | 34.7 | 13.875 | 0.1 |
| Fallow land | 0.651 | 0.424 | 7.7 | 3.928 | 0.1 |
| Land use/land cover | 0.765 | 0.585 | 16.1 | 11.489 | 0.1 |

Four significant factors, accounting for 69.7% of the variation in the original variables were revealed through factor analysis. Factor I reflects the economic characteristics of the households, while factor II is considered an indicator of education and conservation knowledge. Factor III represents the government policy in soil conservation, while factor IV is described as the land use/land cover factor. Stepwise regression shows that forest clearance, fallow land, and land use/land cover are contributing for 16.1% of the variation in the original variables. Stepwise regression indicates that 58 percent of variation in the dependent variables is explained by forest clearance, fallow land, and land use/land cover. The F-values for these predictor variables are significant at 0.1%. The R-squared value (0.585) denotes that about 58 percent variation in the dependent variable (soil erosion rate) is explained by the predictors (forest clearance, fallow land, and land use/land cover) for W. Kufranja watershed.

According to farmers' perceptions, structural measures which have been practiced for a long time are considered the most efficient soil conservation measures adopted to reduce soil erosion. However, to be effective, structural measures by themselves must be integrated with agronomic measures. Contour stone terraces (or stone bunds) are used both on hillslopes, grazing and barren lands for soil and water conservation, and for afforestation purposes in both high and low rainfall conditions. The terraces break the slope and reduce the velocity of surface water. Similarly, check dams are normally constructed on gullies and small ravines formed by erosive activity of water (Figure 9). Ideally, a check dam must be located in a narrow part of a gully with high banks, where it can collect water and sediments, and might be used for cultivation, and tree planting (for afforestation). Such a procedure reduces runoff velocity and erosive activity. Continuous storage of water behind a check dam often improves soil moisture of the adjoining areas, thus improving the natural vegetation cover which in turn increases the cohesion of soil against erosion. However, as reported above soil erosion is not a recent problem for Jordan, it has been present in the highlands of the country since the Iron Age. The destruction of vegetation and the resultant high rates of soil erosion probably date back to the Neolithic and, the Nabateans (≈from 120 BC), may have already lived in a landscape that is very similar to the one existing today. It is obvious that the farmers of W. Kufranja are familiar with soil conservation measures and they have practiced these for a long time. When the government launched the soil conservation project during the 1970s, they obtained the conservation structures installed during the project free of cost. Therefore, at present, some 90% of the farmers are unwilling to pay for or to share in the cost of conservation structures to control soil erosion across the watershed.

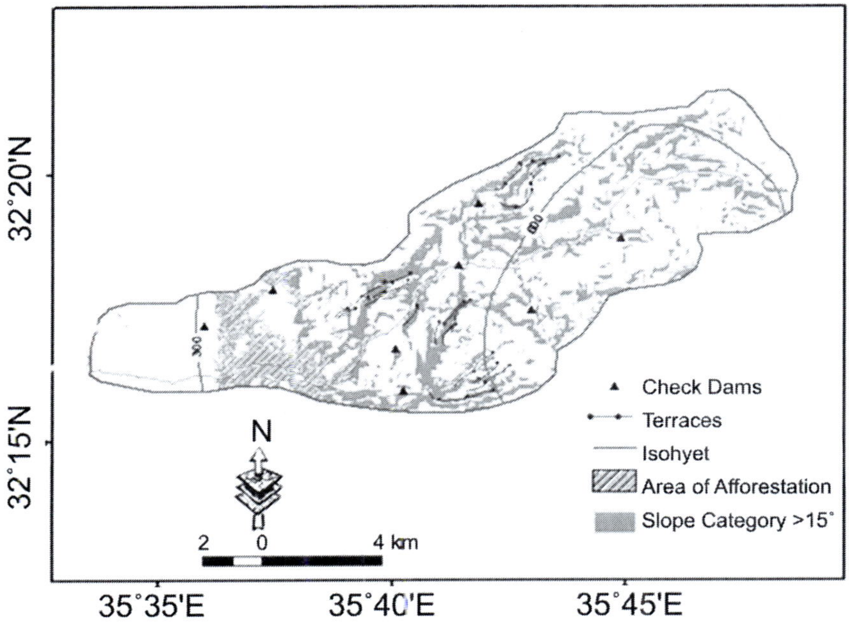


**Figure 9**. Soil conservation measures suggested to control soil erosion in W. Kufranja. (Based on photo interpretation and fieldwork).

In view of such attitudes, the authors suggest that government may offer the farmers soft loans encouraging them to restore and renew the old damaged conservation structures, and to install new structures. Vegetative measures are also recommended by respondents to prevent further forest degradation. In this context, forest plantation could be expanded in the watershed between 100 to 400 m a.s.l (Figure 9). The annual rainfall in this part of the catchment is ≈300 mm, and the slope ranges between 12° and 50°. In 2010, the estimated bare ground across the catchment was 11.9% km$^2$, and about 6.6 km$^2$ of this area was suitable for reforestation. This means that there is an excellent opportunity to reduce surface runoff and soil erosion in this part, which in turn can increase the recharge of sub-surface aquifers.

The farmers' perceptions and socioeconomic determinants of soil erosion in the present case study are based on geospatial techniques (GIS and RS), field measurements of soil erosion, and a household survey. Few significant variables were employed. Thus, it is possible to extend the adopted techniques using large parameters to generate more precise information which can help in formulating more efficient soil conservation plans. A long-term monitoring program for soil erosion estimation, and evaluation of the impact of conservation structures in reducing soil loss and improving farms productivity, are essential in order to achieve sustainable agriculture.

## REFERENCES

1. The Natural Resources Authority (1965) Report on floods in Jordan. Unpublished Report, NRA, Amman.

2. McDonald Partners, and Hunting Technical Services Ltd. (1965) East Bank Water Resources Summary Report. Central Water Authority, Amman.

3. FAO, UNEP, and UNESCO (1979) A Provisional Methodology for Soil Degradatio Assessment. FAO, Rome.

4. Ananzeh, A. (1986) Sediment Yield of Wadi Kufranja Basin. M.A. Thesis, University of Jordan, Amman. (In Arabic)

5. Beni Taha, O. (2004) Soil Erosion in Jerash Area. M.A. Thesis, University of Jordan, Amman. (In Arabic)

6. Khresat, S.A., Al-Bakri, J. and Al-Tahhan, R. (2008) Impact of Land Use/Cover Change on Soil Properties in the Mediterranean Region of Northwestern Jordan. Land Degradation and Development, 19, 397-407.

7. Karadsheh, E.A., Akroush, S. and Mazahreh, S. (2012) Land Degradation in Jordan-Review of Knowledge Resources. International Center for Agricultural Research in the Dry Areas (ICARDA) and USAID, Amman.

8. Khresat, S.A., Rawajfih, Z. and Mohammad, M. (1998) Land Degradation in North-Western Jordan: Causes and Processes. Journal of Arid Environments, 39, 623-629.

9. Al-Zitawi, F. (2006) Using RUSLE in Prediction of Soil Loss for Selected Sites in North and North West of Jordan. MSc Thesis, Jordan University of Science and Technology, Irbid.

10. Farhan, Y., Zreqat, D. and Farhan, I. (2013) Spatial Estimation of Soil Erosion Risk Using RUSLE Approach, RS and GIS Techniques: A Case Study of Kufranja Watershed, Northern Jordan. Journal of Water Resources and Protection, 5, 1247-1261.

11. Farhan, Y. and Nawaiseh, S. (2015) Spatial Assessment of Soil Erosion Risk Using RUSLE and GIS Techniques. Environmental Earth Sciences.

12. Alkharabsheh, A.M., Alexandridis, T.K., Bolas, G., Misopolinos, N. and Silleos, N. (2013) Impact of Land Cover Change on Soil Erosion Hazard in Northern Jordan Using Remote Sensing and GIS. Procedia Environmental Sciences, 19, 912-921.

13. Farhan, Y., Zreqat, D. and Nawaysa, S. (2014) Assessing the Influence of Physical Factors on Spatial Soil Erosion Risk in Northern Jordan. Journal of American Science, 10, 29-39.

14. Lindsay, P.H. and Norman, D.A. (1972) Human Information Processing: An Introduction to Psychology. Academic Press, New York.

15. Dayour, F., Yendaw, E. and Jasaw, G.S. (2015) Local Residents' Perception and Adaptation/Coping Strategies to Climate-Induced Disasters in Bankpama, Wa West District, Ghana. International Journal of Development and Sustainability, 3, 2186-2205.

16. Quennell, A.M. (1958) The Structure and Geomorphic Evolution of the Dead Sea Rift. Quarterly Journal of the Geological Society, 114, 1-24.

17. Fisher, W.H., Atkinson, K., Beaumont, P., Coles, A. and Gilchrist-Shirlaw, D. (1966) Soil Survey of Wadi Ziqlab, Jordan. University of Durham, Durham.

18. Farhan, Y. (1982) Slope Morphology in Central and Southern Jordan. University of Yarmouk Press, Irbid.

19. Farhan, Y. (1986) Landslides in Central Jordan with Special Reference to the March 1983 Rainstorm. Singapore Journal of Tropical Geography, 7, 80-96.

20. Beaumont, P. and Atkinson, K. (1969) Soil Erosion and Conservation in Northern Jordan. Journal of Soil and Water Conservation, 2, 144-147.

21. Fisher, W.H., Bowen-Jones, H., Atkinson, K., Beaumont, P., Smith, K. and Stevens, J. (1969) Soil and Land Potential Survey of the Highlands of North-West Jordan. University of Durham, Durham.

22. Atkinson, K. and Beaumont, P. (1971) The Forests of Jordan. Economic Botany, 25, 305-311.

23. Zregat, D., Alhusban, Y. and Alzamat, K. (2014) Forest Land-Cover Change in Wadi Kufranja Basin/Jordan during the Period 1978-2010: Using Remote Sensing (RS) and Geographic Information System (GIS) Techniques. Muta (A Research Journal Published by the University of Muta, Kerak, Jordan), 29, 125-162.

24. Saket, S.K. (1970) Naur Landslide at km 25. Unpublished Report, Natural Resources Authority, Amman.

25. Saket, S.K. (1975) Slope Instability on the Jordanian Highways. Unpublished Report, Ministry of Public Works, Amman.

26. Farhan, Y. (1988) Problems of Slope Instability on the Amman-Jerusalem Highway: Naur-Adasiya Area, Central Jordan. Studia Geomorphologica Carpatho-Balcanica, 22, 91-104.

27. Farhan, Y. (1988) Hill-Slope Environment and Instability Problems on the Amman-Dead Sea Highway, Central Jordan. Hannon Research Journal (University of Lebanon), 20, 49-72.

28. Farhan, Y. (1999) Landslide Hazards and Highway Engineering in Central and Northern Jordan. In: Griffiths, J.S., Stokes, M.R. and Thomas, R.G., Eds., Landslides-Proceedings of the Ninth International Conference and Field Trip on Landslides in Bristol, UK, A.A. Balkema, Amsterdam, 37-45.

29. Farhan, Y. (2002) Slope Stability Problems in Central and Northern Jordan. The Arab World Geographer, 5, 265-290.

30. Lal, R. (2001) Soil Degradation by Erosion. Land Degradation and Development, 12, 519-539.

31. Wischmeier, W.H. and Smith, D.D. (1965) Predicting Rainfall Erosion from Cropland East of the Rocky Mountains-Guide for Selection of

Practices for Soil and Water Conservation. Agricultural Handbook, No. 282, US Department of Agriculture, Washington DC.

32. Renard, K.G., Foster, G.R., Weesies, G.A. and Porter, J.P. (1991) RUSLE: Revised Universal Soil Loss Equation. Journal of Soil and Water Conservation, 46, 30-33.

33. Renard, K.G. and Freimund, J.R. (1994) Using Monthly Precipitation Data to Estimate the R Factor in the Revised USLE. Journal of Hydrology, 157, 287-306.

34. Renard, K.G., Foster, G.R., Yoder, D.C. and McCool, D.K. (1994) RUSLE Revisited: Status, Questions, Answers and the Future. Journal of Soil and Water Conservation, 49, 213-220.

35. Renard, K.G., Foster, G.R., Weesies, G.A., McCool, D.K. and Yoder, D.C. (1997) Predicting Soil Erosion by Water: A Guide to Conservation Planning with the Revised Universal Soil Loss Equation (RUSLE). Agricultural Handbook No. 703, US Department of Agriculture, UAD-ARS, Washington DC.

36. Meijerink, A.M.J. (1988) Data Acquisition and Data Capture through Terrain Mapping Units. ITC Journal, 7, 23-44.

37. Ministry of Agriculture (1995) The Soil of Jordan: Report of the National Soil Map and Land Use Project, Ministry of Agriculture (Jordan). Hunting Technical Services Ltd, and European Commission, Amman.

38. Morgan, R.P.C. (1977) Soil Erosion: Topics in Applied Geography Series. Longman, London.

39. Morgan, R.P.C. (1977) Soil Erosion in the United Kingdom: Field Studies in the Silsoe Area 1973-1975. Occasional Paper No. 4, National College of Agricultural Engineering, London.

40. Speight, J.G. (1977) Landform Pattern Description from Aerial Photographs. Photogrammetria, 32, 161-182.

41. Hansen, A. (1984) Landslide Hazard Analysis. In: Brunsden, D. and Prior, D.B., Eds., Slope Instability, Wiley, New York, 523-602.

42. van Zuidam, R. (1979) Terrain Analysis and Classification Using Aerial Photographs: A Geomorphological Approach. ITC Text Book of Photo-Interpretation, 1, Enschede.

43. van Zuidam, R. (1985) Aerial Photo-Interpretation in Terrain Analysis and Geomorphological Mapping. Smite, The Hague.

44. Verstappen, H.T. (1983) Applied Geomorphology: Geomorphological Survey for Environmental Development. Elsevier, Amsterdam.

45. Departmrnt of Statistics (DOS) (2010) Statistical Yearbook. Amman, Jordan.

46. Yamani, T. (1967) Statistics: An Introductory Analysis. Harper and Row, New York.

47. Udayakumara, E.P.N., Shrestha, R.P., Samarakoon, L. and Schmidt-Vogt, D. (2010) People's Perception and Socioeconomic Determinants of Soil Erosion: A Case Study of Samanalawewa Watershed, Sri Lanka. International Journal of Sediment Research, 25, 323-339.

48. Adebayo, A.A. (1997) Factor Analysis of Forest Floor Variability in a Savanna Woodland of Southwestern Nigeria. Global Journal of Pure and Applied Science, 3, 255-263.

49. Olorunlana, F.A. (2015) Factor Analysis of Soil Variability in Okoko Region of Ondo State, Nigeria. Journal of Geography and Regional Planning, 8, 12-15.

50. Eltaif, N., Gharaibeh, M., Al-Zaitawi, F. and Alhamad, M. (2010) Approximation of Rainfall Erosivity Factors in North Jordan. Pedosphere, 20, 711-717.

51. Abu Hammad, A. (2011) Watershed Erosion Risk Assessment and Management Utilizing Revised Universal Soil Loss Equation-Geographic Information System in the Mediterranean Environments. Water and Environment Journal, 25, 149-162.

52. Kouli, M., Soupios, P. and Vallianatos, F. (2009) Soil Erosion Prediction Using the Revised Universal Soil Loss Equation (RUSLE) in a GIS Framework, Chania, Norther Crete, Greece. Environmental Geology, 57, 483-497.

53. Alexakis, D., Hadjimitsis, D. and Agapiou, A. (2013) Integrated Use of Remote Sensing, GIS and Precipitation Data for the Assessment of Soil Erosion Rate in the Catchment Area of Yialias in Cyprus. Atmospheric Research, 131, 108-124.

54. Demirci, A. and Karaburun, A. (2012) Estimation of Soil Erosion Using RUSLE in a GIS Framework: A Case Study in the Buyukcekmece Lake Watershed, Northwest Turkey. Environmental Earth Sciences, 66, 903-913.

55. Zhang, X., Wu, B., Ling, F., Zeng, Y., Yan, N. and Yuan, C. (2010) Identification of Priority Areas for Controlling Soil Erosion. Catena, 83, 76-86.

56. Jabbar, M.T. and Chen, X. (2005) Soil Degradation Risk Prediction Integrating RUSLE with Geo-Information Techniques, the Case of Northern Shaanxi Province in China. American Journal of Applied Sciences, 2, 550-556.

57. Wu, X. and Wang, X. (2011) Spatial Influence of Geographical Factors on Soil Erosion in Fuyang County, China. Procedia Environmental Sciences, 10, Part C, 2128-2133.

58. Krishna Bahadur, K.C. (2012) Spatio-Temporal Patterns of Agricultural Expansion and Its Effect on Watershed Degradation: A Case Study from the Mountains of Nepal. Environmental Earth Science, 65, 2063-2077.

59. Christopherson, G. and Guertin, D. (1995) Soil Erosion, Agricultural Intensification, and Iron Age Settlement in the Region of Tall el-Umeiri. Unpuplished Paper Presented at the Annual Meeting of American Schools of Oriental Research, Philadelphia.

60. Cordova, C.E. (1999) Landscape Transformation in the Mediterranean-Steep Transition Zone of Jordan: A Geoarchaeological Approach. The Arab World Geographer, 2, 182-201.

61. Cordova, C.E. (2000) Geomorphological Evidence of Intense Prehistoric Soil Erosion in the Highlands of Central Jordan. Physical Geography, 21, 357-374.

62. Kaliski, A. (1978) A Survey of Soil Erosion and Land Use within the Lower Sector of the Zerqa River Catchment Area. Final Report, No. 3, Amman, Jordan.

63. German Agricultural Technical Cooperation Team (1982) Technical Cooperation with the Hashemite Kingdom of Jordan: A Survey of Soil Erosion and Land Use within the Lower Sector of the Zerqa River Catchment Area. Unpublished Report, Amman, Jordan.

64. Ai-Alawi, M. and Abujamous, M. (2009) Estimation of Soil Erosion in Jordan by Using GIS, In: De Amicis, R., Stojanovic, R. and Conti, G., Eds., Geospatial Visual Analytics: Geographical Information Processing and Visual Analytics for Environmental Security, Springer, Berlin, 439-450.

65. Sharaiha, R. and Ziadat, F. (2007) Alternative Cropping Systems to Control Soil Erosion in Arid to Semi-Arid Areas of Jordan. African Crop Science Conference Proceedings, 8, 1559-1565.

66. Abu-Zreig, M.M., Tamimi, A. and Alazba, A.A. (2011) Soil Erosion Control and Moisture Conservation of Arid Lands with Stone Cover. Arid Land Research and Management, 25, 294-307.

# Chapter 3

# Casting a Wider Net: Understanding the "Root" Causes of Human-Induced Soil Erosion

**Michele A. Whitecraft [1,*] and Bruce E. Huggins Jr.[2]**

[1]Department of Education, Mansfield University, Mansfield, PA 16933, USA

[2]Alderhollow Gardens, South Plymouth, NY 13844, USA

## ABSTRACT

Although science has helped us to identify and measure the threat of soil erosion to food production, we need to cast a wider net for effective solutions. Honest assessment suggests, in fact, that this kind of eco-agricultural issue exceeds the traditional boundaries of scientific interest. The issue of soil erosion spills out so many ways that it demands a holistic interdisciplinary approach. In this paper we explore a systems "in context" approach to understanding soil erosion built upon the interplay of Aristotle's virtues of *episteme*, *techne*, and *phronesis*. We model the synergy of collaboration, where diverse ways of knowing, learning and being in the world can offer proactive soil conservation strategies—those that occur from the inside-out—instead of reactive policies, from the outside-in. We show how positivist scientific attitudes could well impede conservation efforts insofar as they can inhibit educational pedagogies meant to reconnect us to nature. In so doing, we make the ultimate argument that disparate fields of knowledge have much to offer each other and that the true synergy in solutions to soil erosion will come from the intimate interconnectedness of these different ways of knowing, learning and being in the world.

## KEYWORDS

soil erosion; *episteme*; *techne*; *phronesis*; practical wisdom; integrative education; transformative education; experiential education; soil health

## 1. INTRODUCTION

Although science has helped us to identify and measure the threat of soil erosion to food production, we need to cast a wider net for effective solutions (since the world gets 99.7% of its food calories from land and less than 0.3% from the ocean and other aquatic sources [1,2], truly "casting a wider net" is a fitting metaphor here). Honest assessment suggests, in fact, that this kind of eco-agri-cultural issue exceeds the traditional boundaries of scientific interest. The issue of soil erosion spills out so many ways that it demands a holistic interdisciplinary approach.

Accordingly, we, a scientist who is also a teacher and a farmer who is also a scholar, explore a systems "in context" approach to understanding soil erosion built upon the interplay of Aristotle's virtues of *episteme*, *techne*, and *phronesis*—roughly understood as the theoretical "know why" of science, the technical "know how", and the practical knowledge and ethics [3]. Through this paradigm we endeavor to round out the literature and till some practical wisdom, some prudence, into the technical discussion. We also model the synergy of collaboration, where diverse ways of knowing, learning and being in the world can offer proactive strategies—those that occur from the inside-out—instead of reactive policies, from the outside-in. We believe this approach supplies us with soil-conservation solutions superior to those generated through the linear and reductive modes of traditional science, rightfully respected, but not well-matched to our problem. In fact, positivist scientific attitudes could well impede conservation efforts here insofar as they can inhibit educational pedagogies meant to reconnect us to nature. We instead cast beyond science alone in order to foster a more pro-active interdependence and understanding of the foundation of life—soil. In so doing, we make the ultimate argument that disparate fields of knowledge have much to offer each other and that the true synergy in solutions to soil erosion will come from the intimate interconnectedness of these different ways of knowing.

## 1.1. Aristotelian Framing

A scientific consensus holds that overconsumption, population growth, and the use of faulty technologies within inappropriate socio-political-economic arrangements servicing that consumption, are fast destroying humanity's natural capital [4]. Science determines this, but as well, all science is done within these same insufficient arrangements, and so it seems reasonable to incorporate politically and economically attuned sociological knowledge in the study of human-induced problems of soil degradation. Therefore, we chose to frame this paper using Flyvbjerg's seminal piece, Making Social Science Matter [3], where he elaborates on the writings of Aristotle's intellectual virtues outlined in The Nicomachean Ethics. He summarizes the three virtues as follows:

***Episteme***: Scientific knowledge. Universal, invariable, context-independent. Based on general analytical rationality. The original concept is known today from the terms "epistemology" and "epistemic".

***Techne***: Craft/art. Pragmatic, variable, context-dependent. Oriented toward production. Based on practical instrumental rationality governed by a conscious goal. The original concept appears today in terms such as "technique", "technical", and "technology".

***Phronesis***: Ethics. Pragmatic, variable, context dependent. Oriented toward action. Based on practical value-rationality. Deliberation about values with reference to praxis. The original concept has no analogous contemporary term. [Terms that are similar are "applied ethics" or "policy studies"].

As Aristotle noted, working with people is not the same as manipulating experimental apparatuses; choosing practical actions for the human realm requires a different kind of thinking than one employed for making such choices in the mechanical realm. *Techne* is the proper rationality for developing knowledge to determine actions that deal with the physical realm; *phronesis* is the proper rationality for developing knowledge to determine actions (praxis) that deal with people [5]. Science, in this respect, is about *techne* for sure. Human-induced soil degradation is about people, hence *phronesis*. The best solutions to soil erosion in general, we believe, will come from the synergy of the interplay of the three.

## 1.2. Background Terminology

*Episteme*, a seemingly quiet, objective word, is presently a space of much political contention. Though the word "epistemology" is derived from *episteme*, it no longer means only one kind of knowing (*i.e.*, a state of mind capable of demonstrating what is known through the scientific method). Today it is used in a broader sense as in, for example, "feminist epistemology" and "constructivist epistemology." Pressley's ([6], p. 3) description of feminist epistemology exhibits that broader sense and the pressure that it puts onto our sense of knowledge:

Feminist epistemology is concerned with "whose knowledge" is being considered. Feminist epistemologists critique traditional epistemology and argue for ways of understanding knowledge that focus on context and situation. Feminist epistemologists do not suggest that empirical evidence is wrong, but rather that it is necessary to understand that most beliefs are as much a result of their social context as they are factually true. The particulars of knowledge construction are the main focus for feminist epistemologists, rather than universal circumstances for justifying knowledge. These philosophers are often working on undertakings that are political in addition to intellectual.

The kind of knowledge Aristotle calls the *episteme* is what this description refers to as "empirical evidence", which to a researcher is that which is directly observed with the senses in a systematic process of inquiry known as the scientific method [7]. However, in a different context "empirical" can simply mean "experience"; even experiences absent of "controls" and "treatments". In matters that involve the human decision-making scientism falls short. The principles of rational inquiry do not always apply to choices people make, yet that does not mean that they are erroneous or inferior (think about love and altruism; think about dreams so big that facts don't matter). People are different and the universal, context-independent rules of *episteme* are not the proper rationality for assessing the human condition and motivations. So why are scholars so afraid to incorporate other ways of knowing into finding solutions to human-induced problems?

## 2. HISTORICAL PERSPECTIVE

The answer seems likely tagged to cultural context. The cultural framework of western science sets itself up as universal, but it is really carved out in opposition to other ways of knowing the world and underscored by human politics. For instance, in the book, *The Truth About Stories*, King [8] contrasts the Christian story of creation with the Native American story of creation: the elements in Genesis create a particular universe governed by a series of hierarchies—God, man, animals, plants—that celebrate law, order and good government, while our Native American story, the universe is governed by a series of co-operations—animals and humans celebrating equality and balance. He then argues that the Genesis story creates a tone of competiveness, whereas with the Native story the world is at peace, and the pivotal concern is not with the ascendancy of good over evil but with the issue of balance. King ([8], pp. 23–24) continues:

So here are the choices: a world in which creation is solitary, individual act or a world in which creation is a shared activity; world that begins in harmony and slides toward chaos or world that begins in chaos and moves toward harmony; a world marked by competition or a world determined by co-operation. You recognize this pairing is a dichotomy, the elemental structure of Western society. And cranky old Jacques Derrida notwithstanding, we do love our dichotomies. Rich/poor, white/black, strong/weak, right/wrong, culture/nature, male/female, written/oral, civilized/barbaric, success/failure, individual/communal. We trust easy oppositions. We are suspicious of complexities, distrustful of contradictions, fearful of the enigmas.

Not unlike the Genesis act of creation, science is often portrayed as a solitary, individual act [9]. Moreover, formed within this dualistic model, science historically has occupied (or been occupied by) the same side of the duality where King locates western values in general. Pressley ([6], p. 5) echoes King's argument when she discusses the politically tuned consequences of dualistic thinking around which western conceptions of "knowledge" are built. She says:

Western philosophy is built around the idea of binary oppositions such as reason/emotion, mind/body, universal/particular, objective/subjective, and male/female. These are typically hierarchical with the first term given privilege. This dualistic thinking has led to the association of maleness with reason, mind, objectivity, and universals while femaleness is associated with emotion, body, subjectivity, and particulars. Feminist scholars often argue that these dichotomies create one type of knowledge that is masculine. These theorists argue that the period for singular methodology and theory has passed, and it is time to incorporate new standpoints into our way of understanding truth.

Perhaps, ironically, the singular methodology and perceived superiority of the scientific method grew from the stories that have lived in our communal subconscious—stories that smothered our desire for connectedness and communion even as they enacted the social, human element of all our thinking [7]. Perhaps, even if these genesis-type stories were not our own, they are the stories that infused our mainstream science, and what we do as scientists is validated within their paradigm [10].

Regardless, those so inculcated in the scientific method, may feel extremely uncomfortable embracing other ways of knowing (recall the heresy of those who went against the dominant culture and embraced a heliocentric universe). Alternatively, others may find it refreshing as they may intuitively realize that multiple views and different ways of knowing provide the most objective insights into any phenomena as Nietzsche [11] suggests: "There is *only* a perspective seeing, *only* a perspective "knowing"; and the *more* affects we allow to speak about one thing, the *more* eyes, different eyes, we can use to observe one thing, the more complete our "concept" of this thing, our "objectivity", be (emphasis in original)." One can actually see a shift in the *episteme* towards this notion of integrated perspectives play out in the evolution of agricultural practices, to which we next turn.

## 2.1 Toward a More Holistic Approach

In the past, soils were checked for their chemical content and the plant was fed with chemicals [12]. Typically, to offset nutrient losses erosion inflicts on crop production, large quantities of fertilizers were applied [12]. Narrowly considered, it seems like an efficient process, driven by objective numbers. Widen the assessment, however, and it is clearly not so efficient. Those fertilizers are fossil-fuel based and thus caught up in planet-wide competition over these resources; they pollute and disrupt ecosystems; and recent scientific findings strongly suggest that these methods do not even support the ways plants best access nutrients [13,14,15]. Ultimately, this type of agricultural practice is unsustainable. Now, many farmers and researchers are trying to understand the whole picture, which includes caring about the environmental effects and the sources of material input, including the politics, economics, and sociological consequences. It also includes a better understanding of the fungal and bacterial life that is essential to nourishing healthy plant life. Chemical fertilizers do not create these ideal conditions. Instead, we are now looking at more naturally available, biological inputs. For example, farmers can plant a tillage radish (Figure 1) to penetrate compacted soil allowing for better water storage and less run off [16]. Likewise, instead of adding fertilizers to enhance phosphorus we can now work to create the conditions necessary for mycorrhizal fungi to flourish and support phosphorus uptake [13].

**Figure 1.** Left: degraded soil with low organic matter; Right: same soil type with different management resulting in greater soil health. The tillage radish provides a biological method to breakup compaction, add organic matter and store nutrients for the spring. With permission to publish photos from Dorn Cox, University of New Hampshire.

The following diagram (Figure 2) depicts the holistic approach upon which the synergy of science, technology and practical wisdom merge to develop sustainable soil conservation practices.

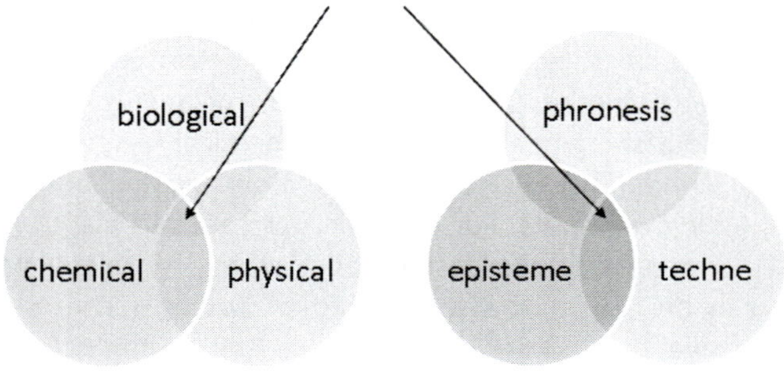


**Figure 2.** Holistic approaches honor the interconnectedness of dynamic systems and enable synergistic solutions.

This holistic approach is part of the Cornell Soil Assessment protocol. Researchers at Cornell have devised an holistic approach that addresses the biological, chemical and physical aspects of soil health, together. On the next page (Table 1), we provide an example of Cornell's suggested management strategies for addressing soil health constraints. Although a holistic approach that assesses the interconnectedness of each component may be more practical with developed countries, developing countries may rely on the same principles through an intimate relationship with an earth whose life-giving properties are treated with respect and appreciation. Despite a lack of *episteme* (know why), or a *techne* (technology to accomplish "know how"), all of us can use the *phronesis* (practical wisdom, ethics, and prudence) that evolves through the intimate experience and relationship with nature. Next we look at how a positivist scientific attitude can impede conservation efforts, how technology in the absence of *phronesis* contributes to the problem and how education interventions involving *phronesis* can re-connect us to nature and foster a more proactive interdependence and understanding of the foundation of life—soil.

## 3. SETBACKS OF POSITIVISM (*EPISTEME*)

Unfortunately, the strong residue of traditional positivist thinking in science can impede conservation efforts. The academy is still in the grip of that gaping and imbalanced *episteme*, even as we look to educational interventions to reconnect us to nature and the foundation of our life in the life of the soil. With a few notable exceptions, the kind of knowledge most valued in the academy is that which can be dealt with in a rational, linear, and controlled manner. These words, "rational", "linear", and "controlled", Palmer claims in *The Heart of Higher Education*, define the comfort zone in academic culture ([17], p. 36). As an academic himself, Palmer goes so far as to say that academia is sometimes governed by orthodoxy as profound as any church. Further, Palmer cites examples where formulaic thinking conceals rather than reveals the truth, and sometimes with catastrophic results. He mentions, among other examples, the complicity of German Higher Education in the Holocaust where the purity obsession forbade "real scholars" from engaging in the messy affairs of the world. Citing Polanyi, Palmer ([17], p. 22) says that our explicit knowing depends on a "vast subterranean layer of tacit human knowing, and we will be arrogant about the hegemony of science until we learn to honor its wordless underground foundations".

**Table 1.** Suggested management strategies for soil health constraints [14].

| | Short Term or Intermittent | Long Term |
|---|---|---|
| **Physical Concerns** | | |
| Low aggregate stability | Fresh organic materials (shallow-rooted cover/rotation crops, manure, green clippings) | Reduced tillage, surface mulch, rotation with sod crops |
| Low available water capacity | Stable organic materials (compost, crop residues high in lignin, biochar) | Reduced tillage, rotation with sod crops |
| High surface density | Limited mechanical soil loosening (e.g., strip tillage, aerators); shallow-rooted cover crops, bio-drilling, fresh organic matter | shallow-rooted cover/rotation crops; avoid traffic on wet soils; controlled traffic |
| High subsurface density | Targeted deep tillage (zone building, etc.); deep rooted cover crops | Avoid plows/disks that create pans; reduced equipment loads/traffic on wet soils |
| **Biological Concerns** | | |
| Low organic matter content | Stable organic matter (compost, crop residues high in lignin, biochar); cover and rotation crops | Reduced tillage, rotation with sod crops |
| Low active carbon | Fresh organic matter (shallow-rooted cover/rotation crops, manure, green clippings) | Reduced tillage, rotation |
| Low mineralizable N (Low PMN) | N-rich organic matter (leguminous cover crops, manure, green clippings) | Cover crops, manure, rotations with forage legume sod crop, reduced tillage |
| High root rot rating | Disease-suppressive cover crops, disease breaking rotations | Disease-suppressive cover crops, disease breaking rotations, IPM practices |

| | | |
|---|---|---|
| Chemical concerns | See also soil fertility recommendations | |
| Unfavorable pH | Liming materials or acidifier (such as sulfur) | Repeated applications based on soil tests |
| Low P, K and Minor elements | Fertilizer, manure, compost, P-mining cover crops, mycorrhizae promotion | Application of P, K materials based on soil tests; increased application of sources of organic matter; reduced tillage |
| High salinity | Subsurface drainage and leaching | Reduced irrigation rates, low-salinity water source, water table management |
| High sodium content | Gypsum, subsurface drainage, and leaching | Reduced irrigation rates, water table management |

We might well heed the warning. Significantly, it comes from within science too. Advocating a further break, from our complete reliance on science as linear, objective and reductionist, Zajonc ([18], p. 81) cites Einstein and quantum mechanics:

Einstein's relativity and quantum mechanics both undermine objectification and support a *relational view of reality* in which phenomena are co-created by the observer and the world. Second, through entanglement and emergence, physics offers evidence for the ontological holism that grants wholes a standing long denied them. Parts are no longer privileged. These two realizations are essential to a proper philosophical infrastructure for higher education.

When it comes to soil degradation, agriculture and food scarcity, blindly following the narrow rules of a narrowly defined efficiency without more focus on resilience, sustainability, biodiversity and distributive fairness [19] will not produce sustainable results. In the same vein, the activities of financial participants in the food system have driven commodity prices away from levels justified by "market fundamentals", as the literature shows that food price speculation, not supply and demand, has played the biggest role in excessive price hikes [20,21,22,23]. Further supporting this Zajonc ([15], p. 81) writes:

As in physics, the simplifying assumptions of classical economics were made because economic theory could not handle the complexities of the real world. But are humans really rational economic actors? Economics experiments show we are not. Understanding this opens us to other important questions to consider, such as, is market behavior the only or best way to gauge preferences, or might we allow for thoughtful, patient introspection concerning the root causes of suffering and happiness? Does the market really offer an accurate and comprehensive valuation of community or might we allow for forms of fellowship that elude economic objectification?

According to Zajonc, fields of science need to re-conceive themselves according to a postreductionist paradigm in which lived experience, connection and complexity are given far more attention in the same way that economics has had to rethink its paradigms for understanding human behavior in the market [18]. Palmer ([17], p. 41) corroborates this and notes an ironic connection to pedagogy, that fundamental root zone of the academy:

Academic culture needs to embrace the simple fact that cognition, which is our business, is intimately linked to affect, no matter how much we think emotions are *not* our business. As neuroscientists such as Candice Pert have told us, thinking is not done solely by the brain, an organ housed in the cranium. Thinking is done by the mind, which is not an organ but a process that is distributed throughout the body and draws on every faculty we have. So when I hear faculty dismiss the affective dimension of teaching and learning as "touchy-feely stuff", I have to conclude that they are projecting their personal discomfort with emotions rather than making a statement about the real world.

There is a pair of tragicomic ironies embedded in academic resistance to taking seriously the connection between feeling and thinking. Academics who want to factor out "subjective emotions" in favor of data based "objective knowledge" will, at the same time, blithely ignore fifty years of research about the importance of attending to the emotions if we want to liberate the mind. I do not think it unfair to say that such people are "pedagogical fundamentalists" who proof-text the research the way Biblical fundamentalists proof-text the Bible, honoring whatever supports their biases and ignoring the rest.

The paired irony is that these academics ignore all the research-based knowledge we have on the role of emotions in learning largely because embracing the implications of that knowledge would take them out of their emotional comfort zones (italics in original)! It is enough to make one's head spin. But we who advocate for integrative education ought to be spinning our heads in public, weaving a sound defense for attending to the heart-mind connection, making it more difficult for orthodox academics to be dismissive of brain science, pedagogical reality, and simple common sense.

We wholeheartedly agree that a more holistic, integrative approach to soil erosion is necessary as a complete reliance on science alone has been a major setback in soil conservation.

## 4. "FAULTY" TECHNOLOGY (*TECHNE*)

As documented in "The Plow that Broke the Plains [24]", U.S. farmers (whose tools may have been considered universal and context independent), stayed with the traditional methods and technologies of tillage and decimated the Midwest during the 1930s. Similarly, if we blindly pull through our world with the traditional linear reductionist plow, without concern for system dynamics and the synergy of the interconnected whole, solutions to worldwide soil degradation will be inadequate. Currently our models are at odds with one another which further demonstrates the inadequacies of knowing all of the variables and hence, the characterization of and possible solutions to soil erosion. Consider, for example, the research Trimble [25] cites on U.S. soil erosion alone. In 2000, he cites the average annual cropland soil erosion at 2 billion [26], 4.0 billion [27,28,29], 4.5 billion [30], 4.8 billion [31], 5 billion [32], or 6.8 billion tons. Clearly, the range of predictions makes it difficult to use these models as tools, let alone understand the magnitude and interconnectedness of the variables involved. Trimble [25] does not mention that he is citing models from various decades, there is still great variation within a decade. Nonetheless, scientists do agree that although erosion rates in the U.S. have decreased in the past two decades [2], cropland is still losing soil faster than its sustainable replacement rate [33,34]. Additionally, erosion rates on U.S. rangelands remain relatively high at 6 t/ha-year [35].

In short the models and technologies we use to understand soil erosion and soil degradation do not honor the truly complex, interconnected nature of the problem. We need to go beyond our traditional methodologies and incorporate what Brooks [36] calls an "epistemological modesty". This modest disposition begins with the recognition that there is no one method for solving problems [36]. Relying on quantitative and rational analysis gives you part of the truth, not the whole [36]. Brooks ([36], p. 246), gives this example which leads with *episteme* and *techne*, and ends with *phronesis*:

If you were asked what day in the spring you should plant corn, you could consult a scientist. You could calculate weather patterns, consult the historical record, and find the optimal temperature range and date at each latitude and altitude. On the other hand you could ask a farmer. Folk wisdom in North America decrees that corn should be planted when oak leaves are the size of a squirrel's ear. Whatever the weather in any particular year, this rule will guide the farmer to the right date.

This knowledge is a different sort of knowledge. This knowledge comes from "integrating and synthesizing diverse dynamics and is produced over time, by an intelligence that is associational—observing closely, imagining loosely, comparing like to unlike and like to like to find harmonies and rhythms in the unfolding of events ([36], p. 246)". This associational intelligence can even help advance scientific knowledge, too. Honoring different ways of knowing, honoring the practitioners' intuitions and observations, can also help scientists enrich their research questions. Case in point, one particular farmer [37] has noticed for decades, that the height of a hornets' nest has served as an excellent predictor of the amount of snowfall in a particular area in upstate New York. Instead of a complete reliance on meteorology for weather prediction, might we now study animals to enhance our ability predict the weather?

In short, technology is only as good as the hand that guides it—and the guiding hand is only as wise and compassionate as the mind and heart that direct it [17]. We look next at the practical knowledge, wisdom and ethics that would have prevented, "The Plow that Broke the Plains".

## 5. THE PRUDENCE OF *PHRONESIS*

*"An ethical action, rightly taken, invites reciprocal generosity that can appear in unexpected ways."*

*—Parker Palmer*

The Midwestern soil crisis of the 1930s could have been prevented, had we turned to practical wisdom and ethics, that is, had we looked at the world as a living organism (hence *phronesis*) as opposed to a machine (hence *techne*). We believe that the soil, as well as the world, should be treated as a living organism. Just as we are not simply a compilation of various cells, but a whole that is entirely different than its parts, soil is not just dirt, but a living breathing entity full of organic matter, bacteria, minerals, water and air whose whole is so much greater than the sum of its parts. We understand Barbara McClintock's Nobel winning mentality as described by Palmer ([17], p. 28):

Barbara McClintock understood her primary experimental materials not as objects but as beings. We can know a relational reality only by being in relation to it—not keeping our distance, as in the objectivist mythology, but moving close and leaning in, then testing what we know against the standard of evidence and logic in the context of the scientific community. McClintock, says one writer, "gained valuable knowledge by empathizing with her corn plants, submerging herself in their world and dissolving the boundary between object and observer."

In order for practical knowledge and wisdom to guide our actions and inform our solutions, we must realize that the "objectivity" that has dominated higher education is a myth. We now understand that there is a mutually influential relationship of the knower and the known [38] and that objectivism is no longer a viable way to frame knowing, teaching or learning, let alone policy or decision-making.

Regardless of the field (agricultural or disciplinary), as humans, we must hold *phronesis* at least on equal footing as*episteme* and *techne* if we truly wish to live in a sustainable way. The practical knowledge and wisdom of practitioners must be valued. This knowledge and wisdom, borne of experience and immersion in our various fields, can truly enhance decision-making and problem solving. Consider the veteran teacher, who, in an effort to maximize learning, chooses to review a test at the end of a class instead of the beginning so that those who may have done poorly are not distracted [39]. Or, the veteran police officer who knows just what questions to ask. Case in point, when a female victim of an armed robbery was asked for a description of the burglar, she described the thief (whose face she could not see) as a "man about 5 foot 2 inches tall, wearing camouflage from head to toe [40]". Upon hearing this description, the officer had the foresight to ask what the thief did with the pocketbook. The victim said that "he put it over his shoulder and ran away". As she spoke she began to realize what she was saying and they both concluded that the culprit may have been female; since most men would just hold the purse in their hands and run away. Sure enough, the suspect, a female, was caught twenty minutes later with the camouflage clothing hidden in her car. In this case the preconceived notion that robbers are male, unknowingly added details to the description of the thief that could have inhibited the thief's apprehension. The experience and practical knowledge possessed by the officer allowed his interest to be piqued by the short stature of the culprit, and thereby formulate a question that exposed the true sex of the suspect even when the victim's brain muddled her description of what actually transpired. As cases like the above show, even before something is confirmed as "knowledge" it is still useful.

## 6. EDUCATING "RUNNERS" FOR A MORE TRANSFORMATIVE WORLD

The (not these) United States is (not are) a collective society that thinks as an individualistic one [36]. In one study, Nisbett [41] found that when asked to describe their day, American six year olds make three times more references to themselves than Chinese six year olds. Additionally, the latest trend in U.S. education has been "student-centered" learning as opposed to "teacher-centered". We disagree with the "it's all about me" attitude and subscribe to Palmer's philosophy that teaching and learning should be "subject-centered" [42] as well as experiential to nurture an empathetic, communal understanding of the interconnectedness of life. Through entering into a subject-centered, experiential education we believe that people will have greater opportunities to practice trust; a trust that may lead to compassion and sacrifice. Here is how Brooks ([36], p. 155) describes it:

Trust is habitual reciprocity that becomes coated by emotion. It grows when two people begin volleys of communication and cooperation and slowly learn they can rely upon each other. Soon members of a trusting relationship become willing to not only cooperate with each other but sacrifice for each other.

A closer look at the foundations of American education shows a "hidden curriculum" that is atomistic and competitive rather than interconnected and communal [17]. Earlier we explored Hill's suggestion whereby this philosophy of education may have its underpinnings in the stories we've been told. Regardless, we acknowledge the need for an integrative education that aims to "'think the world together' rather than 'think it apart'; to know the world in a way that empowers educated people to act on behalf of wholeness rather than fragmentation [17]". Like Palmer and Zajonc, we believe in a profound form of education where caring and knowing are united. In an education based on relationships where Goleman [43] finds we are "hard wired for social learning, imitating down to the level of neurons what is occurring around us". In addition, according to Zajonc [18], such imitation is formative, shaping the very cognitive, emotional, and moral faculties we use to make sense of the world. We believe in an education where we:

Attend to the cultivation of our students' humanity at least as much as we instruct them in the content of our fields. Long after they forget the content they learned, who they have become will endure and determine much of the character and quality of their contribution to society and personal satisfaction they take in life ([18], p. 102).

When we consider, that the global urban population in 2008 was at 3.3 billion people [44], more than half of all humans living on the planet, we recognize the urgency that now, more than ever, we need to create opportunities for people to have a life-sustaining relationship with the earth. This statistic stands in stark contrast to 13 percent a century ago; and 3 percent a century before that [44]. The solutions to soil degradation must come from an educated populace that learns from the interrelationship and our interconnectedness to one another and the earth, regardless of whether we live in the city or the country. Compassionate action is fostered in students when they learn not only with the intellect but also with the heart. Transformative, experiential educational opportunities must activate our feelings. For when our feelings are activated, we are moved to action [18].

Unfortunately, the technology of today can act as an obstacle to developing interpersonal skills and relationships as well as life skills and compassion in general. In the past children had to resolve many of their own conflicts as their parents were not instantly accessible (via cell phone) to help navigate a disagreement with a friend. They learned from their successes and their failures. They learned from experience. Today, many U.S. students have been reported surfing the internet and texting during class, bullying each other on social media, "googling" an answer instead of formulating one on their own, and taking pictures of the board as opposed to taking notes [39]. Additionally the advent of "all-online" degrees, suggest we have forgotten about the power of those person-to-person, face-to-face, "live encounters" that animate the human spirit in ways that nothing else can [17]. The kind of "distance education" (both spatial and affective) that objectivism breeds lays the ground for lives lived at a distance from the suffering world [17].

Conversely, technology can bring people and ideas closer together. People who live in remote locations or parents who stay home with their children, now have access to higher education and information they would not have had otherwise. Very simply, a basic tenet of toxicology—the dose makes the poison—is applicable here. What dictates whether technology will have a positive or negative effect on the human and earthly condition centers on practical wisdom and ethics. Practical wisdom—the wisdom associated with experience and thus, always, with specificity, locality—we can cause to be learned through collaborative interactions with one another and with nature. Currently, in the traditional teaching/learning paradigm we teach students *about* nature instead of *from* nature. We need to esteem experiential learning to transform our world.

To bring this paper full circle, we revisit the Native American wisdom in a letter written to the white man:

We know that you highly esteem the kind of learning taught in those Colleges, and that the Maintenance of our young Men, while with you, would be very expensive to you. We are convinced that you mean to do us Good by your Proposal: and we thank you heartily, But you, who are wise must know that different Nations have different Conceptions of things and you will therefore not take amiss, if our ideas of this kind of Education happen not to be the same as yours. We have had some Experience of it. Several of our young People were formerly brought up at the Colleges of the Northern Province: they were instructed in all your Sciences; but, when they came back to us, they were bad Runners, ignorant of every means of living in the woods, neither fit for Hunters, Warriors, nor Counselors, they were totally good for nothing.

We are, however, not the less oblig'd by your kind offer, tho' we decline accepting it; and, to show our grateful Sense of it, if the Gentlemen of Virginia will send us a Dozen of their Sons, we will take Care of their Education, instruct them in all we know and make Men of them ([45], p. 240).

Not only did we find this letter very well crafted for what it says, we thought the wisdom in the carefully chosen words and in what it doesn't say were beautiful examples of *phronesis*. We doubt the white-man's letter would have been as culturally sensitive.

# 7. CONCLUSIONS

Many have called for farmer incentives to fully implement conservation methods to help save the earth's soil [2,46,47]. It is our hope that one day the understanding that we are part of a greater, interconnected whole will be incentive enough to want to do the right in conserving all of the earth's natural resources. Living in a sustainable way, as we have attempted to elucidate in this paper, certainly goes well beyond pecuniary incentives. Putting more money into incentives for farmers to implement conservation methods, is akin to putting more synthetic fertilizer on fields without looking at the interconnectedness of the reasons for the poor quality soil. This is why this paper sought to incorporate a more holistic understanding that incorporates how we educate our populace.

To paraphrase Albert Einstein, you cannot solve a problem with the frame of mind that created the problem. We feel the best solutions to preventing soil degradation will not come from *episteme*, *techne* or *phronesis* alone, but from integration of all three of Aristotle's intellectual virtues, *including* the non-falsifiable one. The richness of the world will not reveal itself by a single means of inquiry [18]. As Palmer [17] describes it, "genuine knowing comes out of a healthy dance between the objective and the subjective, between the analytic and the integrative between the experimental and what I would call the receptive". Some pioneers have realized this and are now beginning to offer solutions to soil erosion by looking at the interconnectedness of the chemical, physical and biological aspects of soil. Likewise we believe that the sensitivity necessary to produce wisdom, we can cause to be learned through collaborative, integrative, and experiential learning.

In the vocabulary of competition, if we continue to value knowledge over understanding, in the end soil will win. Nature—not *episteme* or *techne*—will be the final arbiter. Through *phronesis* we understand that. The life that springs out of soil will also return to it, us too. We just have to decide what legacy we want to leave behind, and if that legacy includes a sustainable lifestyle [48]. After all, soil erosion, soil degradation, and food scarcity are all related to epistemology:

Every epistemology—rooted (as all of them are) in a particular ontology, and manifesting (as all of them do) in a particular pedagogy—has an impact on the ethical formation of learners. Epistemology becomes operational in students' lives not through overt conversation or explicit knowing but through modes of teaching and learning that tacitly form or deform learners in a particular way of relating to the world. An integrative pedagogy is more likely to lead to moral engagement because it engages more of the learners self and teaches by means of engagement: the curriculum *and* the "hidden curriculum" embedded in such a pedagogy support a way of knowing that involves much if not all of the whole self in learning about the world [italics in original] ([17], p. 32).

## ACKNOWLEDGMENTS

We wish to thank: Elaine Norris for her editorial contributions and critique of our paper; Dorn Cox from Tuckaway Farm and doctoral student at University of New Hampshire for use of his tillage radish photo; and the Cornell Soil Health Team for publishing a holistic manual on soil health.

## REFERENCES

1. Food and Agriculture Organization of the United Nations (FAO), *Food Balance Sheet*; FAO: Rome, Italy, 1998.

2. Pimentel, D. Soil erosion: A food and environmental threat. *Environ. Dev. Sustain.* **2006**, *8*, 119–137.

3. Flyvbjerg, B. *Making Social Science Matter: Why Social Inquiry Fails and How It Can Succeed Again*; Cambridge University Press: Cambridge, UK, 2001.

4. Ehrlich, P. Human natures, nature conservation, and environmental ethics. *BioScience* **2002**, *52*, 31–43.

5. Polkinghorne, D. *Practice and the Human Sciences: The Case for a Judgment-Based Practice of Care*; State University of New York Press: Albany, NY, USA, 2004.

6. Pressley, L. *Whose Knowledge Is It Anyway? Feminist Epistemology and Science*; University of North Carolina: Greensboro, NC, USA, 2005.

7. Whitecraft, M. Understanding the Acculturation of Women in Science: The Interplay of *Episteme*, *Techne*, and *Phronesis*in Retaining Females

in Undergraduate Science. Ph.D. Thesis, Cornell University, Ithaca, NY, USA, 15 December 2012.

8. King, T. *The Truth about Stories: A Native Narrative*; University of Minnesota Press: Minneapolis, MN, USA, 2003.

9. Brotman, J.; Moore, F. Girls and science: A review of four themes in the science education literature. *J. Res. Sci. Teach.***2008**, *45*, 971–1002.

10. Norris, E. Personal communication, Syracuse University: Syracuse, NY, USA, 2013.

11. Nietzsche, F. *On the Genealogy of Morals*; Vintage Books: New York, NY, USA, 1969.

12. Moebius-Clune, B. Soil as a Foundation for Health and Resilience. In Presented at Cornell University, Ithaca, NY, USA, 30 April 2013.

13. Wolfe, D. Soil as a Foundation for Health and Resilience. In Presented at Cornell University, Ithaca, NY, USA, 30 April 2013.

14. Gugino, B.; Idowu, O.; Schindelbeck, R.; van Es, H.; Wolfe, D.; Moebius-Clune, B.; Thies, J.; Abawi, G. *Cornell Soil Health Assessment Training Manual*, 2nd ed.; NYSAES: Geneva, NY, USA, 2009.

15. Van Noordwijk, M.; Cadisch, G. Access and excess problems in plant nutrition. *Plant Soil* **2002**, *247*, 25–39.

16. Cox, D. Soil as a Foundation for Health and Resilience. In Presented at Cornell University, Ithaca, NY, USA, 30 April 2013.

17. Palmer, P. *The Heart of Higher Education*; Jossey-Bass: San Francisco, CA, USA, 2010.

18. Zajonc, A. Attending to Interconnection: Living the Lesson. In *The Heart of Higher Educaction*; Jossey-Bass: Sanfrancisco, CA, USA, 2010; pp. 77–99.

19. Paul, J.; Wahlberg, K. A New Era of World Hunger? Available online: http://www.globalpolicy.org/social-and-economic-policy/world-hunger/48424.html (accessed on 10 October 2011).

20. Knaup, H.; Schiess, M.; Seith, A. Speculating with Lives: How Global Investors Make Money out of Hunger. Available online: http://www.globalpolicy.org/component/content/article/217-hunger/50673-speculating-with-lives-how-global-investors-make-money-out-of-hunger.html (accessed on 18 October 2011).

21. Global Policy Forum/Friedrich Ebert Foundation. Interactive Thematic Dialogue of the UN General Assembly on the Global. Available online: http://www.globalpolicy.org/images/pdfs/0406gadialogue.pdf (accessed on 5 October 2011).

22. Food and Agriculture Organization. The State of Food Insecurity in the World 2010. Available online: http://www.fao.org/docrep /013/i1683e/i1683e.pdf (accessed on 3 October 2011).

23. Organization for Economic Cooperation and Development & FAO. Agriculture: Higher Prices Here to Stay, says OCED–FAO. Available online: http://www.globalpolicy.org/social-and-economic-policy/world-hunger/general-analysis-on-hunger/50354-agriculture-higher-prices-here-to-stay-says-oecd-fao-report-.html?itemid=id#1291 (accessed on 10 October 2011).

24. Lorentz, P. The Plow that Broke the Plains. [Film]. Eastern Service Studios: New York, NY, USA, 1936. Available online: http:// www.youtube.com /watch?v=fQCwhjWNcH8 (accessed on 18 May 2013).

25. Trimble, S. U.S. soil erosion rates—Myth and reality. *Science* **2000**, *289*, 248–250.

26. USDA, *Soil and Water Resources Coservation Act: 1980 Appraisal, Part I: Soil, Water, and Related Resources in the United States: Status, Conditions and Trends and Appraisal; Part II: Soil, Water and Related Resources in the United States, Analysis of Trends*; USDA: Washington, DC, USA, 1981.

27. Raven, P.; Berg, L.; Johnson, G. *Environment*; Saunders: Philadelphia, PA, USA, 1995.

28. Beasley, R. *Erosion and Sediment Control*; Iowa State University Press: Ames, IA, USA, 1972.

29. Government Accounting Office, *To Protect Tomorrow's Food Supply, Soil Conservation Needs Priority*; GAO: Washington, DC, USA, 1977.

30. Pimentel, D.; Harvey, C.; Resosudarmo, P.; Sinclair, K.; McNair, M.; Crist, S.; Spritz, L.; Fitton, L.; Saffouri, R.; Blair, R. Response to Crosson letter. *Science* **1995**, *268*, 464–465.

31. Barlow, T. *Soil Conservation Policies: An Assessment*; Soil Conservation Society of America: Akeny, IA, USA, 1979.

32. Pimentel, D. Land degradation: Effects on food and energy resources. *Science* **1976**, *194*, 149–155.

33. USDA, *Changes in Average Annual Soil Erosion by Water on Cropland and CRP Land, 1992–1997*; Natural Resources Conservation Service: Washington, DC, USA, 2000.

34. USDA, *Changes in Average Annual Soil Erosion by Wind on Cropland and CRP Land, 1992–1997*; Natural Resources Conservation Service: Washington, DC, USA, 2000.

35. Committee on Opportunities in Agriculture, National Research Council, *Frontiers in Agricultural Research: Food, Health, Environment and Communities*; National Academies Press: National Academies Press, 2003.

36. Brooks, D. *The Social Animal*; Random House: New York, NY, USA, 2011.

37. Huggins, B. Personal Communication, Alderhollow Gardens: South Plymouth, NY, USA, 2013.

38. Palmer, P. *To Know as We Are Known*; HarperCollins: New York, NY, USA, 1993.

39. Whitecraft, M. Personal Communication, Cornell University: Ithaca, NY, USA, 2013.

40. National Geographic, *Investigation Discovery*; National Geographic Channel: Washington, DC, USA, 2012.

41. Nisbett, R. *The Geography of Thought: How Asians and Westerners Think Differently... and Why*; Free Press: New York, NY, USA, 2003; pp. 87–88.

42. Palmer, P. *The Courage to Teach: Exploring the Inner Landscape of a Teacher's Life*; Jossey-Bass: San Francisco, CA, USA, 1998.

43. Goleman, D. *Social Intelligence: The New Science of Human Relationships*; Bantam: New York, NY, USA, 2006.

44. Ruble, B. The Challenges of the Twenty-First Century City. In Presented at Fetzer-Wilson Center Seminar, Community Resilience in the Twenty-First Century, Chicago, IL, USA, 14 April 2010.

45. Carroll, A. *Letters of a Nation*; Broadway Books: New York, NY, USA, 1999; p. 240.

46. Department of Natural Resources. Soil and Water Conservation Program. Available online: http://dnr.mo.gov/env/swcp/ (accessed on 20 May 2013).

47. United States Government Accounting Office, *Agricultural Conservation: USDA Needs to Better Ensure Protection of Highly Erodible Cropland and Wetlands, Report to the Ranking Democratic Member, Committee on Agriculture, Nutrition, and Forestry; U.S. Senate*; U.S. General Accounting Office: Washington, DC, USA, 2003.

48. Pimentel, D.; Whitecraft, M.; Scott, Z.R.; Zhao, L.X.; Satkiewicz, P.; Scott, T.J.; Phillips, J.; Szimak, D.; Singh, G.; Gonzalez, D.; *et al.* Will limited land, water, and energy, control human population numbers in the future? *Hum. Ecol.***2010**, *38*, 599–611.

# Chapter 4

# Soil Erosion and Its Relationship to the Spatial Distribution of Land Use Patterns in the Lancang River Watershed, Yunnan Province, China

**Yueqiang Li, Shanghong Zhang*, Yang Peng**

Renewable Energy School, North China Electric Power University, Beijing, China

## ABSTRACT

The Lancang River Watershed is one of the most biologically diverse areas in the world. The river flows through Yunnan Province, China, which suffered serious deforestation since the 1980s; this in turn led to increased soil erosion in the region. To investigate the influence of the spatial distribution of land use and slope on soil erosion in the Lancang River Watershed, the Soil and Water Assessment Tool (SWAT) model was used to establish hydrological models using two-phase land use maps (1975 and 1985), a soil map, and meteorological data from 11 gauging stations. The satisfactory values of Nash-Sutcliffe efficiency $E_{ns}$ and correlation coefficient $R^2$ during the calibration and validation period indicated that SWAT can be used in this area to simulate the average annual soil erosion under different land use scenarios change. By comparing soil erosion rate under different land use scenarios change, forests and grasslands had similar effects on preventing soil erosion. A parameter, soil erosion increment ($E_i$), was used to assess the effects of slope on soil erosion. The results revealed that variation in sediment yield was more sensitive to land use change for slopes exceeded 25° than for slopes being 0° - 15°. The spatial distribution of land use also had a relationship to soil erosion. Compared with the soil erosion rate in each sub-watershed using two-phase land use maps, the soil erosion rate increased when the percent cover of natural vegetation decreased. The results of this study provide baseline data for soil conservation and protection of the environment and ecology of the Lancang River Watershed in Yunnan Province.

## KEYWORDS

Lancang River Watershed, Land Use, Slope, Soil Erosion, SWAT, Forest, Grassland

## 1. INTRODUCTION

Soil erosion, a major environmental problem, often leads to loss of nutrients, a decline in soil fertility, and a reduction in crop productivity. Soil erosion is a type of non-point source pollution that can cause river silting and water pollution, and affects normal flood discharge as well as service life and efficiency of hydropower projects [1]. Many factors influence soil erosion, including natural conditions and human activities [2] [3]. Natural factors include geomorphology, soil type, climate, and vegetation. Human activity factors mainly include construction and engineering projects as well as land development and use. Land use change is one of the largest single threats causing increased sediment yield and soil erosion [4] - [8]. During the past several decades, research related to this issue has received increasing attention by scholars globally [9] - [16].

The Lancang River is one of six major water systems in Yunnan Province; human activities in the watershed have exacerbated soil erosion problems [17] [18]. The upstream portion of the Lancang River Watershed has a high percentage of forest cover, but income from timber harvest that creates deforestation has long been the pillar of the local economy. The middle reaches of this watershed experience intensive human disturbance, because highway construction, mining, and power plant construction are concentrated in this area. As a result, this area experiences wide variations in land use, which creates environmental issues. The expanding economy also creates problems with immigration while making sustainable economic development very problematic. The lower reaches of the Lancang River Watershed in the Xishuangbanna district, a tropical area of China, are inhabited by mountain people who have applied traditional slash and burn methods to make use of the land, so rapid changes are occurring in land use and vegetation coverage. In general, the Lancang River Watershed has a very fragile environment that is a result of deforestation and an unsustainable agricultural system [17]. Yao et al. used geographic information system (GIS) to study the spatial distribution of soil erosion and factors influencing soil erosion in the Lancang River Watershed [19], while the study revealed the effect of land use on soil erosion without excluding slope as a factor, and analyzed the response of soil erosion to slope without excluding the acreage factor. Therefore, delineating the areas that are most vulnerable to soil erosion as well as revealing the relationship between soil erosion and the spatial distribution of land use can assist local land use managers in applying appropriate in the Lancang River Watershed.

In recent years, many watershed scale models have been developed to study the hydrological processes [20] - [23]. The Soil and Water Assessment Tool (SWAT) is one of the most suitable models for simulating water and sediment yields underland use and management scenarios [24] - [26]. SWAT as a physically based, distributed, continuous daily time step parameter model, is designed to predict the impact of land management practices on water, sediment and agricultural chemical yields in large, complex watersheds with varying soils, land use and management conditions over long periods of time. Therefore, in this study, the SWAT model was chosen as the tool to study the influence of slope and the spatial distribution of land use on soil erosion.

The main study objectives of this paper were to: (a) explore the effect of land use change and slope on soil ero- sion characteristics in the Lancang River Watershed, Yunnan Province, and (b) analyze the average annual soil ero- sion in each sub-watershed and reveal its relationship to the characteristics of the spatial distribution of land use.

## 2. STUDY AREA

The Lancang River of Yunnan Province drains a 76,147 km$^2$ watershed that covers 23.08% of the Yunnan Province and is located at 21°08' - 29°15'N

98°36' - 102°19'E (see Figure 1); further south it is known as the Mekong River. The long and narrow, north to south aligned watershed has high elevations in the north and low elevations in the south. The main stream of the Lancang River in Yunnan Province, which consists of 96 rivers, is 1240 km long. The climates changes from north to south in the watershed. This is reflected in the average annual temperature that increases from 4.8°C to 21.1°C when going from high to low latitudes and elevations. The entire watershed covers seven climatic zones: cold temperate zone, temperate zone, warm temperate zone, north subtropics, mid-subtropics, south subtropics, and north tropical zone. Most regions of the Lancang Watershed receive an average of more than 1000 mm precipitation annually; most rain (~85%) falls between May and October due to the southwest monsoon. The main land use types in this watershed are forest and farmland.

# 3. MATERIALS AND METHODS

## 3.1. Data Collection

The digital elevation model (DEM), which consists of a sampled array of elevations for a number of ground positions at regularly spaced intervals, land use map, soil map, and climatic data used to drive the SWAT model for this study were collected as follows (1) DEM data with a spatial resolution of 30 × 30 m was downloaded from the International Scientific Data Service Platform (http://datamirror.csdb.cn). (2) Two 1:100,000 land use maps, one each from 1986 and 2000, were obtained from the Data Sharing Infrastructure of Earth System Science (www.geodata.cn), and land use patterns were then classified into six types: forest, grass, river, urban, paddy field, and upland field. In this paper, the paddy field and upland field were collectively known as grain field. (3) The spatial distribution of soil types (1:1,000,000) and soil data based on soil physical characteristics data collected from the Environment and Ecological Science Data Center for West China (http://westdc.westgis.ac.cn). The classification standard of soil particles used in the SWAT model is the standard system used in the United States and the data base used in this study applied the same standard, so soil particle data did not need to be converted. (4) The daily rainfall, maximum and minimum air temperature, wind speed, and humidity data at 11 gauges collected during 1970-2000 was downloaded from the China Meteorological Data Sharing Service System (http://cdc.cma.gov.cn). Because of the lack of measured solar radiation data during the model running time, a weather generator was established to simulate solar radiation. The monthly climatic statistical data which were used in the weather generator was calculated based on the measured daily data from 1951-1973 of Weixi, Baoshan, Jinghong and Lincang gauge station. (5) The day runoff and sediment yield data collected at Jinghong Gauging Station during 1975-1985 were taken from a book named The Hydrological Data of South Tibetan and West Yunnan [27].

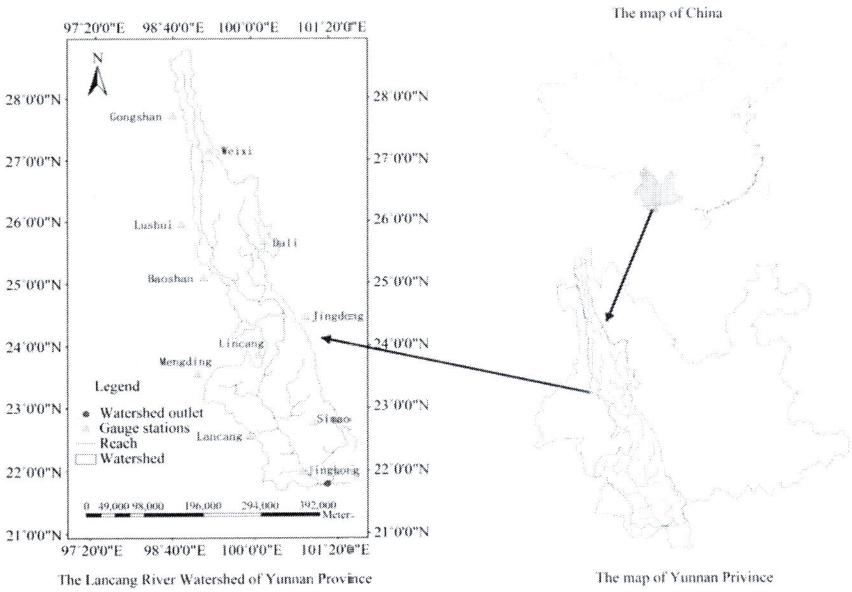


**Figure 1**. Location of Lancang River watershed in Yunnan province, China.

## 3.2. Model Setup

SWAT (Soil & Water Assessment Tool) is a river basin scale model developed to quantify the impact of land management practices in large, complex watersheds. SWAT is a public domain model actively supported by the USDA Agricultural Research Service at the Grassland, Soil and Water Research Laboratory in Temple, Texas, USA [24] - [26]. To use the SWAT model, the watershed is first divided according to the watershed area threshold and the outlet of the watershed. When using a DEM to extract the river network in this study, the threshold value was set as 50,000 ha and the outlet was selected at the position where the Lancang River flows out of Yunnan Province. Finally, the entire watershed was divided into 71 sub-watersheds. An HRU is the basic unit of the SWAT model simulation; this study applied the multiple land use/soil-type method and set the HRU threshold as 10/10, namely if the percentage of one kind land use or soil type is less than 10%, this land use or soil type would be neglectful and replaced by adjacent type; the entire Lancang Watershed, as defined above, was divided into 1705 HRUs.

In this study, the simulation of runoff applied the SCS runoff curve number method; the skewed normal distribution method was adopted to simulate the rainfall; the potential evaporation was calculated based on the Penman-Monteith method and the channel calculus applied the variable storage method which had high simulation precision. More details could be found in theoretical documentation [28].

The purpose of parameter sensitivity analysis is to discover how the variation of the input parameters influences the output and to determine which input parameter variation has the most influence on the output; that is, the sensitivity analysis results will provide a reference for determining the most sensitive parameters. This study compared the simulated runoff and sediment yield with measured data, and sensitivity analysis was used to determine the importance of several parameters that affected the accuracy of runoff and sediment yield simulation results. Appropriate parameters were selected (Table 1 and Table 2) and the model was calibrated accordingly.

## 3.3. Evaluation Indicator

Model performance was evaluated by the correlation coefficient $R^2$ and the Nash-Sutcliffe efficiency $E_{ns}$ [29]. The two indicators can be computed by the following equations:

$$R^2 = \left[ \frac{\Sigma\left(Q_m - \overline{Q_m}\right)\left(Q_s - \overline{Q_s}\right)}{\sqrt{\Sigma\left(Q_m - \overline{Q_m}\right)^2}\sqrt{\Sigma\left(Q_s - \overline{Q_s}\right)^2}} \right]^2 \quad (1)$$

$$E_{ns} = 1 - \frac{\Sigma\left(Q_m - Q_s\right)^2}{\Sigma\left(Q_m - \overline{Q_m}\right)^2} \quad (2)$$

where $Q_m$ and $Q_s$ are the observed and simulated data, and $\overline{Q_m}$ are the $\overline{Q_s}$ mean values of the observed and simulated data. If $R^2 > 0.6$ and $E_{ns} > 0.5$, the model performance can be regarded as acceptable. The higher the values, the better the model performance.

**Table 1**. Calibrated values for runoff simulation.

| The most sensitive parameters[a] | Parameter definition | Range | Calibrated value |
|---|---|---|---|
| $CN_2$ | Initial SCS CN II value | ±25% | −16%[b] |
| SOL_AWC | Soil available water capacity | ±30% | −10%[b] |
| ALPHA_BF | Base flow alpha factor | ±30% | −12%[b] |
| ESCO | Soil evaporation compensation factor | 0.01 - 1 | 0.95 |
| $CH\_K_2$ | Channel effective hydraulic conductivity | 0 - 100 | 53 |

a. Parameters are ordered according to their level of sensitivity; b. Indicates the percentage by which the original values changed.

**Table 2**. Calibrated values for sediment yield simulatio.

| The most sensitive paramters[a] | Parameter definition | Range | Calibrated value |
|---|---|---|---|
| SPCON | Linear parameters for sediment re-entrainment | 0.0001 - 0.01 | 0.0011 |
| SPEXP | Exponent parameter for sediment re-entrainment | 0.8 - 2.0 | 0.9 |

a. Parameters are ordered according to their level of sensitivity.

## 3.4. Study the Effects of Land Use Change and Slope on Soil Erosion

Based on Chinese hydraulic erosion intensity classification reference indicators, reclaiming fields with slopes of 15° - 25° would probably lead to moderate erosion that harms the growth of crops. Reclaiming fields with slopes greater than 25° would lead to severe erosion, and steeply sloping fields are the main source of water and soil loss. Therefore, based on the practical situation in the study area, seven different vegetation recovery scenarios were established to simulate the impact of land use change on sediment yield (see Table 3); other land use-type areas remained unchanged in the simulation. By comparing the soil erosion rate under different scenarios, one can see the effect of land use change on soil erosion in the Lancang River Watershed. Two main types of scenarios were developed: converting grain fields to either forest or grassland. This was done because 63.7% of the entire watershed is forested, while grassland and farmland (including paddy fields and upland fields) account for 18.4% and 16.4% of the landscape, respectively. Also, farming activities are the main element influencing vegetation cover. Therefore, this study emphasized the analysis of the effect of returning grain fields to forest or grassland on soil erosion in the Lancang River Watershed.

Slope is also a major factor influencing sediment yield and soil erosion. To determine the effect of slope on soil erosion rate, a parameter named soil erosion increment was defined using Equation (3).

$$E_i = \frac{|S_i - S_1|}{A_i} \tag{3}$$

where $E_i$ is soil erosion increment (t/km$^2$), $S_1$ and $S_i$ are the sediment yield of land use scenario 1 and scenario i (t), $S_1$ is the reference scenario and $A_i$ is the area of farmland changed during scenario i. The soil erosion increment provides the relative variation in sediment yield per unit area so that a larger $E_i$ indicates greater variation. Comparing the soil erosion increment under different land use scenarios can reveal whether the variation of sediment yield is sensitive to the variations in slope [30]. In this part of the study, the land use of 1986 and the average annual rainfall were used to simulate the soil erosion values.

## 3.5. Assessing the Effects of the Spatial Distribution of Land Use on Soil Erosion

Changes in the spatial distribution of land use could change the characteristics of soil erosion, so the effects of the spatial distribution of land use on soil erosion had to be assessed. This study simulated the average annual soil erosion rate in each sub-watershed using two-phase land use maps. Five sub-watersheds with the most serious variation in soil erosion were selected for analysis. The soil erosion rate and land use types in these sub- watersheds were compared for 1986 and 2000 to clearly identify the effects of land use change on soil erosion. This method removed the effects of the intensity of precipitation, slope, and soil conditions on the soil erosion so that the results only showed the effects of the spatial distribution of land use on soil erosion in the sub- watershed scale. This part of the study helped to identify the areas that were most vulnerable to soil erosion and revealed the influence of land use change in a practical way. It will also help land use managers to conduct land use planning more efficiently.

# 4. RESULTS AND DISCUSSION

## 4.1. Runoff and Sediment Yield Calibration and Validation

Data collected on runoff and sediment yield from 1975 to 1985 and meteorological data from 1970 to 2000 were used in this study. The years of 1970-1974 were selected as the warm-up period, 1975-1980 were the calibration year, and 1981-1985 were the validation year. Figure 2 and Figure 3 compare measured and simulated runoff and sediment yield at the Jinghong Gauging Station during the calibration and validation years.

**Table 3**. Calibrated values for runoff simulation.

| Scenarios | 1 | 2 | 3 | 4 | 5 | 6 | 7 |
|---|---|---|---|---|---|---|---|
| The sloping fields of 0° - 5° | Original condition | Return grain field to forest | Return grain field to grass | - | - | - | - |
| The sloping fields of 15° - 25° | Original condition | - | - | Return grain field to forest | Return grain field to grass | - | - |
| The sloping fields over 25° | Original condition | - | - | - | - | Return grain field to forest | Return grain field to grass |

Table 4 presents the simulation performance for runoff and sediment yield. The $E_{ns}$ values were more than 0.5 and $R^2$ values were generally more than 0.7 for both the calibration and validation years, so one can conclude that the SWAT model can be used to study the impact of the spatial distribution of land use on characteristics of soil erosion in the Lancang River Watershed.

## 4.2. Effects of Land Use Change and Slope on the Soil Erosion

In this study watershed, grain field with slopes of 0° - 15°, 15° - 25° and over 25° accounting for 61.10%, 30.79% and 8.11%, respectively. The average annual soil erosion rate (t/km$^2$) under land use scenarios, 1, 2, 3, 4, 5, 6 and 7 were 1152, 879, 891, 815, 835, 448 and 510, respectively. The results showed that the soil erosion rate decreased when grain fields were returned to forest or grassland. This indicates that both forestland and grassland are in favour of reducing soil erosion when compared with grain fields. And returning grain fields to forest had a slightly better effect on reducing erosion than returning grain fields to grassland.

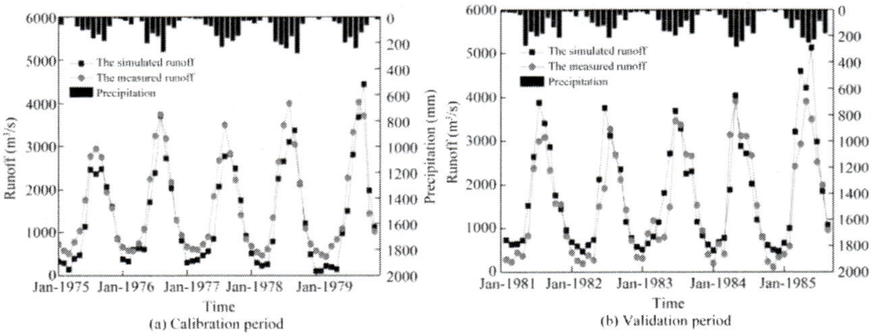


**Figure 2**. Measured and simulated runoff for the calibration and validation periods.

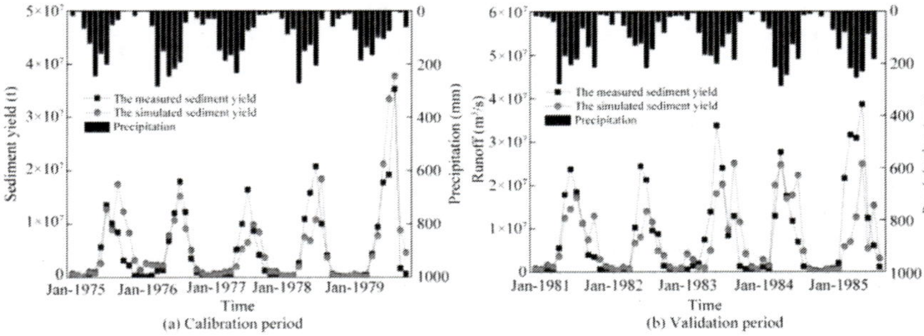


**Figure 3**. Measured and simulated sediment for the calibration and validation periods.

**Table 4**. Simulation performances for runoff and sediment yield.

| Period | Runoff | | Sediment yield | |
|---|---|---|---|---|
| | $E_{NS}$ | $R^2$ | $E_{NS}$ | $R^2$ |
| Calibration | 0.85 | 0.91 | 0.68 | 0.73 |
| Validation | 0.79 | 0.82 | 0.51 | 0.72 |

Table 5 shows $E_6$ is 10,892 t/km$^2$, which is 2.2 times the value of $E_2$. This indicates that the variation of sediment yield is more sensitive to land use change when the slope exceeds 25˚; the same conclusion can be drawn from a comparison of $E_7$ and $E_3$ because $E_7$ is 2.1 times the value of $E_3$. From an economic perspective, this result shows that managing 1 km$^2$ of farmland with a slope over 25˚ was equal to managing almost 2 km$^2$ of farmland with a slope of 0˚ - 15˚. Land managers should seriously consider changes in land use for farmland with slopes exceeding 25˚ to achieve very efficient land use planning and management. The radio of $E_6$ to $E_2$ in this study area is different with the radio in other watershed although they are all bigger than 2, the rainfall intensity and soil type are the main influencing factors, and the different radios between different watersheds need further study. In addition, the fact that $E_2 > E_3$, $E_4 > E_5$ and $E_6 > E_7$ (Figure 4) illustrates that the variation of sediment yield is slightly sensitive to forest than grassland. Because the difference was not obvious, returning grain fields to grassland would be a more cost-effective method than returning grain fields to forest.

## 4.3. Effect of the Spatial Distribution of Land Use on Soil Erosion

Figure 5 provides the soil erosion under average annual rainfall in each sub-watershed under the two-phase land use maps. In Figure 5(a) and Figure 5(b) the five colors represent the different soil erosion levels based on the erosion hazard classification established by the Ministry of Water Resources of the People's Republic of China [31]: mired, mild, medium, intense, and very intense (Table 6). One can conclude soil erosion in downstream areas is greater than that in the upstream and midstream areas because high-intensity precipitation mainly occurs in the downstream areas of the watershed. To quantify the effects of the spatial distribution of land use on soil erosion, five sub-watersheds (25, 26, 36, 38 and 59) were selected because these five sub-watersheds obviously experienced the most variation in soil erosion from 1986 to 2000 (Figure 5(c)). By comparing the two land use maps, it can be seen that all five of these sub-watersheds have obvious land use changes from 1986 to 2000. The forest and grasslands were collectively classified as natural vegetation. FromTable 7 one can conclude that four of the five sub-watersheds (excluding 36) experienced an increase in soil erosion from 1986 to 2000. These changes in soil erosion were mainly caused by a decrease in natural vegetation coverage in the four sub-water- sheds and an increase of natural vegetation in sub-watershed 36. There is a negative proportional relationship between the soil erosion rate and natural vegetation, indicating an increase in the removal of natural vegetation results in a greater amount of soil erosion. These results also indicate that under the same intensity of precipitation and the same slope and soil conditions, that forest and grassland both have the effect on reducing soil erosion when compared with farmland. Local policy makers could base their decisions on protecting the most vulnerable areas as they change landscape use patterns in an attempt to realize sustainable development of water and soil resources in the Lancang River Watershed of Yunnan Province.

**Table 5**. The soil erosion increment of different land use scenarios.

| | | | | | | | |
|---|---|---|---|---|---|---|---|
| Sediment yield ($10^5$ t) | $S_1$ | $S_2$ | $S_3$ | $S_4$ | $S_5$ | $S_6$ | $S_7$ |
| | 873.76 | 666.83 | 675.31 | 618.21 | 632.84 | 339.55 | 386.43 |
| Area ($km^2$) | | $A_2$ | $A_3$ | $A_4$ | $A_5$ | $A_6$ | $A_7$ |
| | | 4153.75 | 4153.75 | 2661.5 | 2661.5 | 4904.5 | 4904.5 |
| Soil erosion increment ($t/km^2$) | | $E_2$ | $E_3$ | $E_4$ | $E_5$ | $E_6$ | $E_7$ |
| | | 5054 | 4778 | 9602 | 9052 | 10892 | 9936 |

# 5. CONCLUSIONS

The SWAT model was used to simulate average annual sediment yield in the Lancang River Watershed, Yunnan, China under different land use and slope scenarios to investigate the effects of the spatial distribution characteristics of land use and slope on soil erosion.

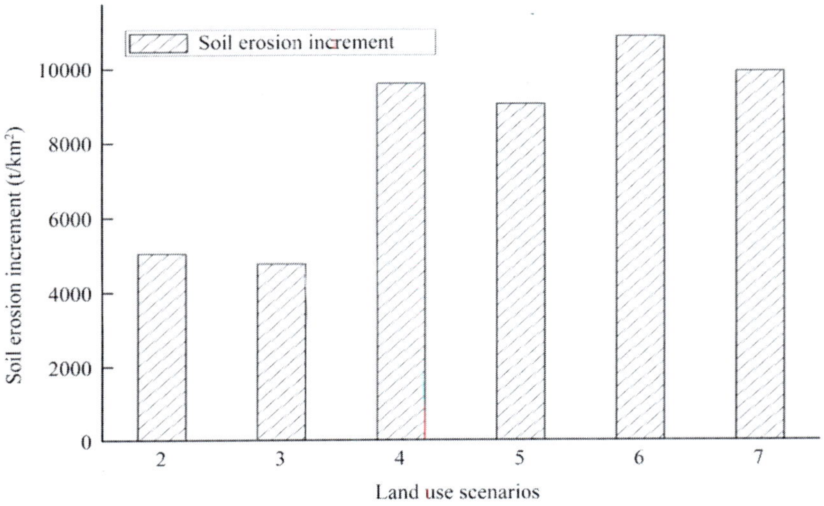


**Figure 4**. The soil erosion increment under different land use scenarios.

By comparing soil erosion rate under different land use scenarios, both forests and grasslands are seen to have similar effect on reducing soil erosion. A parameter named soil erosion increment ($E_i$) was defined to assess the effect of slope on soil erosion. Farmland with slopes over 25° was seen to be more sensitive to land use change in terms of increased relative erosion rate than farmland with slopes of 0° - 15° and 15° - 25°, the $E_i$ ratio between the area over 25° and 0° - 15° is above 2, and forests were slightly better than grasslands in reducing erosion on steep slopes when land use changed.

The characteristics of the spatial distribution of the soil erosion rate under average annual rainfall in each sub- watershed were analyzed for 1986 and 2000 and the area in the middle and lower reaches of the river along the mainstream channel experienced serious soil erosion. Based on the comparison of the five sub-watersheds, a negative proportional relationship exists between the soil erosion rate and forest area. If less land is covered by natural vegetation, the soil erosion rate would obviously increase.

The findings of this study can provide baseline information and reference data in support of soil conservation and environmental protection efforts in the Lancang River Watershed.

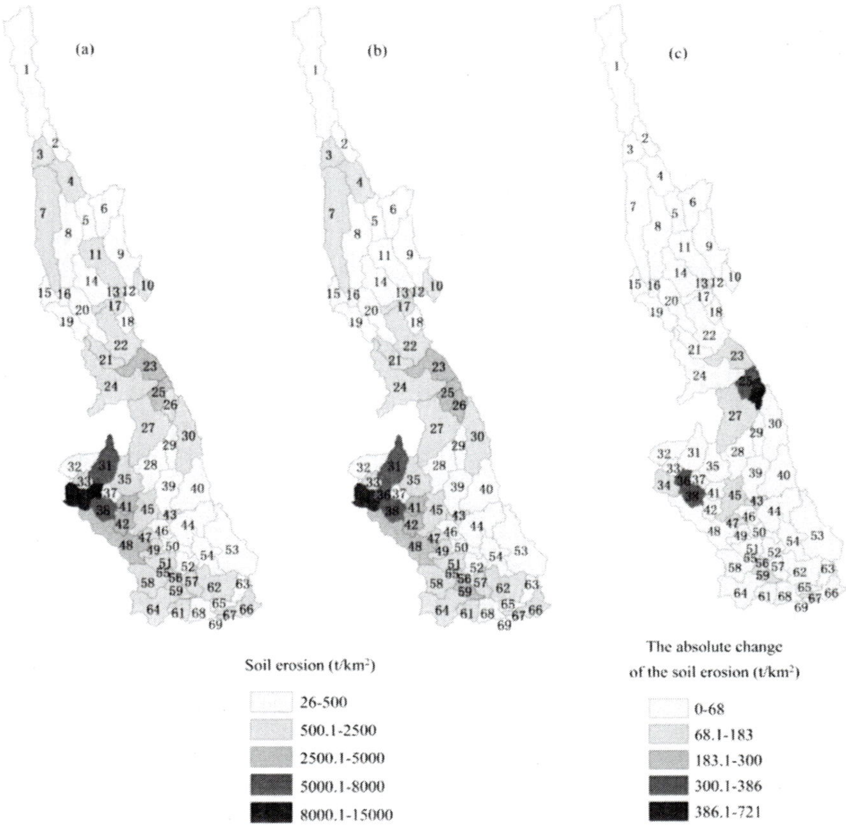


**Figure 5**. Spatial distribution of the average annual soil erosion rate for the two-phase land use map: (a) 1986, (b) 2000, (c) the absolute change in soil erosion in each sub-watershed.

**Table 6**. Classification of erosion intensity.

| Level | Mired | Mild | Medium | Intense | Very intense |
|---|---|---|---|---|---|
| Average erosion modulus (t/(km$^2$·a)) | <500 | 500 - 2500 | 2500 - 5000 | 5000 - 8000 | 8000 - 15,000 |

**Table 7**. The main land use changes and concurrent changes in the soil erosion rate in five sub-watersheds.

| Sub-watershed | Land use types | Area of each land use type ($km^2$) | | Variation ($km^2$) | Soil erosion rate($t/km^2$) | | Variation ($t/km^2$) |
|---|---|---|---|---|---|---|---|
| | | 1986 year | 2000 year | | 1986 year | 2000 year | |
| | Forest | 418.5 | 415.25 | −3.25 | | | |
| 25 | Grass | 104.25 | 91.5 | −12.75 | 4473.3 | 4859.8 | +386.5 |
| | Sum | 522.75 | 506.75 | −16 | | | |
| | Forest | 450.25 | 444.25 | −6 | | | |
| 26 | Grass | 38.5 | 27 | −11.5 | 1916.4 | 2637.7 | +721.3 |
| | Sum | 488.75 | 471.25 | −17.5 | | | |
| | Forest | 102.5 | 99.5 | −3 | | | |
| 36 | Grass | 182.25 | 186.25 | +4 | 8057.8 | 7705.5 | −352.3 |
| | Sum | 284.75 | 285.75 | +1 | | | |
| | Forest | 418 | 395.5 | −22.5 | | | |
| 38 | Grass | 103 | 66.25 | −36.75 | 5749.3 | 6052.6 | +303.3 |
| | Sum | 521 | 461.75 | −59.25 | | | |
| | Forest | 380.25 | 363 | −17.25 | | | |
| 59 | Grass | 34.5 | 47.25 | +12.75 | 2321.2 | 2504.7 | +183.5 |
| | Sum | 414.75 | 410.25 | −4.5 | | | |


## ACKNOWLEDGEMENTS

This study has been supported by the National Natural Science Foundation of China (51379076), the Fundamental Research Funds for the Central Universities and the 12th Five-Year National Key Technology R&D Program (2012BAB05B05). The authors wish to thank: the International Scientific & Technical Data Mirror Site, Computer Network Information Center, Chinese Academy of Sciences for providing DEM data, the Data Sharing Network of Earth System Science, which provided land use maps, the Environmental and Ecological Science Data Center for West China, the National Natural Science Foundation of China, which provided data for soil physical characteristics and the China Meteorological Data Sharing Service System, which supports the collection of meteorological data.

# REFERENCES

1. Wilson, G.V., Cullum, R.F. and R&oumlmkens, M.J.M. (2008) Ephemeral Gully Erosion by Preferential Flow through a Discontinuous Soil-Pipe. Catena, 73, 98-106.

2. Hessel, R. and Jetten, V. (2007) Suitability of Transport Equations in Modelling Soil Erosion for a Small Loess Plateau Catchment. Engineering Geology, 91, 56-71.

3. Vrieling, A., Jong, S.M., Sterk, G. and Rodrigue, S.C. (2009) Timing of Erosion and Satellite Data: A Multi-Resolution Approach to Soil Erosion Risk Mapping. International Journal of Applied Earth Observation and Geoinformation, 10, 267-281.

4. Marques, M.J., Bienes, R., Jiménez, L. and Pérez-Rodríguez, R. (2007) Effect of Vegetal Cover on Runoff and Soil Erosion under Light Intensity Events. Rainfall Simulation over USLE Plots. Science of The Total Environment, 378, 161-165.

5. Bakker, M.M., Govers, G., van Doorn, A., Quetier, F., Chouvardas, D. and Rounsevell, M. (2008) The Response of Soil Erosion and Sediment Export to Land-Use Change in Four Areas of Europe: The Importance of Landscape Pattern. Geomorphology, 98, 213-226.

6. García-Ruiz, J.M. (2010) The Effects of Land Uses on Soil Erosion in Spain: A Review. Catena, 81, 1-11.

7. Feng, X., Wang, Y., Chen, L., Fu, B. and Bai, G. (2010) Modeling Soil Erosion and Its Response to Land-Use Change in Hilly Catchments of the Chinese Loess Plateau. Geomorphology, 118, 239-248.

8. Shen, Z., Chen, L., Hong, Q., Qiu, J., Xie, H. and Liu, R. (2013) Assessment of Nitrogen and Phosphorus Loads and Causal Factors from Different Land Use and Soil Types in the Three Gorges Reservoir Area. Science of the Total Environment, 454-455, 383-392.

9. Hao, F.H., Cheng, L.Q., Liu, C.M. and Dai, D. (2004) Impact of Land Use Change on Runoff and Sediment Yield. Journal of Soil and Water Conservation, 18, 5-8.

10. Ward, J.P., Balen, R.T., Verstraeten, G., Renssen, H. and Vandenberghe, J. (2009) The Impact of Land Use and Climate Change on Late Holocene and Future Suspended Sediment Yield of the Meuse Catchment. Geomorphology, 103, 389-400.

11. Nunes, A.N., Almeida, A.C. and Coelho, C.A.O. (2011) Impacts of Land Use and Cover Type on Runoff and Soil Erosion in a Marginal Area of Portugal. Applied Geography, 31, 687-699.

12. Blavet, D., Noni, G.D., Bissonnais, Y.L., Leonard, M., Maillo, L., Laurent, J.Y., et al. (2009) Effect of Land Use and Management on the Early Stages of Soil Water Erosion in French Mediterranean Vineyards. Soil and Tillage Research, 106, 124-136.

13. Cai, T., Li, Q.F., Yu, M.X., Lu, G.B., Cheng, L.P. and Wei, X. (2012) Investigation into the Impacts of Land-Use Change on Sediment Yield Characteristics in the Upper Huaihe River Basin, China. The Physics and Chemistry of the Earth, 53-54, 1-9.

14. Ouyang, W., Skidmore, A.K., Hao, F. and Wang, T. (2010) Soil Erosion Dynamics Response to Landscape Pattern. Science of the Total Environment, 408, 1358-1366.

15. Ouyang, W., Hao, F.H., Skidmore, A.K. and Toxopeus, A.G. (2012) Soil Erosion and Sediment Yield and Their Relationships with Vegetation Cover in Upper Stream of the Yellow River. Science of the Total Environment, 409, 396-403.

16. Wu, Y. and Chen, J. (2012) Modeling of Soil Erosion and Sediment Transport in the East River Basin in Southern China. Science of the Total Environment, 441, 159-168.

17. Xu, J.C., Zhang, P.F. and Wang, Y.H. (2003) Land Use and Land Cover in Lancang Watershed of Yunnan. Acta Botanica Yunnanica, 25, 145-154. (In Chinese)

18. Dong, S.C., Zhou, C.J. and Wang, H.Y. (2002) Ecological Crisis and Countermeasures of the Three Rivers' Headstream Regions. Journal of Nature Resources, 17, 713-719. (In Chinese)

19. Yao, H.R., Yang, Z.F. and Cui, B.S. (2006) Spatial Analysis on Soil Erosion of Lancang River Watershed in Yunnan Province under the Support of GIS. Geographical Research, 25, 421-425. (In Chinese)

20. Chen, H., Xu, C.Y. and Guo, S.L. (2012) Comparison and Evaluation of Multiple GCMs, Statistical Downscaling and Hydrological Models in the Study of Climate Change Impacts on Runoff. Journal of Hydrology, 434-435, 36-45.

21. Jones, R.N., Chiew, F.H.S., Boughton, W.C. and Zhang, L. (2006) Estimating the Sensitivity of Mean Annual Runoff to Climate Change Using Selected Hydrological Models. Advances in Water Resources, 29, 1419-1429.

22. Du, J.K., Qian, L., Rui, H.Y., Zuo, T.H., Zheng, D.P., Xu, Y.P. and Xu, C.Y. (2012) Assessing the Effects of Urbanization on Annual Runoff and Flood Events Using an Integrated Hydrological Modeling System for Qinhuai River Basin, China. Journal of Hydrology, 464-465, 127-139.

23. 23. Beskow, S., Mello, C.R., Norton, L.D., Curi, N., Viola, M.R. and Avanzi, J.C. (2009) Soil Erosion Prediction in the Grande River Basin, Brazil Using Distributed Modeling. Catena, 79, 49-59.

24. Romanowicz, A.A., Vanclooster, M., Rounsevell, M. and Junesse, I.L. (2005) Sensitivity of the SWAT Model to the Soil and Land Use Data Parametrisation: A Case Study in the Thyle Catchment, Belgium. Ecological Modelling, 187, 27-39.

25. Asres, M.T. and Awulachew, S.B. (2010) SWAT Based Runoff and Sediment Yield Modelling: A Case Study of the Gumera Watershed in the Blue Nile Basin. Ecohydrology & Hydrobiology, 10, 2-4.

26. Baker, T.J. and Miller, S.N. (2013) Using the Soil and Water Assessment Tool (SWAT) to Assess Land Use Impact on Water Resources in an East African Watershed. Journal of Hydrology, 486, 100-111.

27. Hydrological Stations in Yunnan Province (1985) The Hydrological Data of South Tibetan and West Yunnan. Hydrological Stations in Yunnan Province Publisher, Yunnan Province.

28. Neitsch, S.L., Arnold, J.G., Kiniry, J.R. and Williams, J.R. (2005) Soil and Water Assessment Tool (SWAT) Theoretical Documentation. Blackland Research Center, Texas Agricultural Experiment Station and Grassland, Soil and Water Research Laboratory, Temple, TX.

29. Nash, J.E. and Sutcliffe, J.V. (1970) River Flow Forecasting through Conceptual Models Part I—A Discussion of Principles. Journal of Hydrology, 10, 282-290.

30. Zhang, S.H., Liu, Y. and Wang, T.W. (2014) How Land Use Change Contributes to Reducing Soil Erosion in the Jialing River Basin, China. Agricultural Water Management, 133, 65-73.

31. Ministry of Water Resources of the People's Republic of China (2008) Standards for Classification and Gradation of Soil Erosion. SL190-2007.

# Chapter 5

# Uncovering the Footprints of Erosion by On-Farm Maize Cultivation in A Hilly Tropical Landscape

**Chaminda Egodawatta [1], Peter Stamp [2,*] and Ravi Sangakkara [3]**

[1]Department of Plant Sciences, Faculty of Agriculture, Rajarata University of Sri Lanka, Anuradhapura 50000, Sri Lanka

[2]Institute of Agricultural Sciences, ETH Zurich, 8092 Zurich, Switzerland

[3]Department of Crop Science, Faculty of Agriculture, University of Peradeniya, Peradeniya 20400, Sri Lanka

## ABSTRACT

A hilly region in Sri Lanka was considered to be degraded by erosion driven by intensive tobacco production, but what are reliable indicators of erosion? In addition to determining soil chemical and physical traits, maize was cropped with Nitrogen, Phosphorus and Potassium (NPK, PK) recommended mineral fertilization and without fertilizer (ZERO) in two major seasons(October–January in 2007/2008 and 2008/2009—Seasons 1 and 2 respectively) on 92 farms at inclinations ranging from 0% to 65%. In a subset of steep farms ($n$ = 21) an A horizon of 6 cm rather than of 26 cm was strong proof of erosion above 30% inclination. Below the A level, the thickness of the horizon was unaffected by inclination. Soil organic matter contents (SOM) were generally low, more so at higher inclinations, probably due to greater erosion than at lower inclination. Maize yields decreased gradually with increasing inclination; at ZERO, effects of climate and soil moisture on yield were easier determined and were probably due to long-term erosion. However, despite an initial set of 119 farms, an exact metric classification of erosion was impossible. NPK strongly boosted yield. This was a positive sign that the deficits in chemical soil fertility were overriding physical soil weaknesses. The study illustrated that chemical soil fertility in these soils is easily amenable to modifications by mineral and organic manures.

## KEYWORDS

degraded soils; soil nutrients; inclination positions; mineral fertilizers

## 1. INTRODUCTION

Soil erosion is an important widespread type of land degradation, which poses severe limitations on the sustainable use of agricultural land [1]. Degradation of the soil is the main cause of declining productivity on inclined arable lands in the tropics, especially in South and Southeast Asia [2]; not including irrigated terraces, lands in hilly areas constitutes 60% to 90% of the arable land in South Asia [3]. Tropical hillside areas are populated by smallholders, who generally practice subsistence agriculture. Driven primarily by socioeconomic constraints, land use and land cover determine the structure, functioning and dynamics of these landscapes [4]. In tropical hilly regions, erosion is primarily due to water (high-intensity rainfall) and is associated with the removal of soil-protective vegetation, leading to semi-intensified but unsustainable land use.

The rate of soil degradation depends on the rate of land-cover degradation, which in turn is influenced both by adverse climatic conditions and changes in land management. Removal of vegetation means a decrease in the protection of the soil and increases surface runoff and sediment yields, resulting in soil degradation [5]. Due to depletion of organic matter, productivity of the system declines because of the degraded soil structure, a depletion of nutrients and organic matter. On the other hand, extensive rainfed cultivation of crops is restricted mainly to hilly lands with shallow soil, which is prone to erosion under traditional systems [6]. Although steep slopes are usually classified as unsuitable for the continuous production of arable crops, population pressure generally leads to more intensive utilization despite low yields. To counteract this, it is possible to intensify agricultural production by mineral fertilizers and more suitable farm management [7]. Though this may minimize soil degradation due to erosion [8], farmers must be certain that these investments will bring satisfactory returns.

Soil organic matter (SOM) plays an important role as a pool of terrestrial carbon [9], in maintaining the productivity and sustainability of agricultural and non-agricultural ecosystems. A decrease in soil organic matter leads to a reduction in the cation exchange capacity (CEC) [10], resulting in poor nutrient [11] and water [6] retention. Therefore, loss of SOM is usually correlated with significant nutrient depletion [12,13]. A decrease in organic matter leads to a significant decrease in the availability of micro-nutrients such as Zn, Cu, Mn and Fe [14]. Therefore, long-term fertilizer experiments can establish relationships between soil organic matter and soil fertility and the management of organic matter, showing which patterns of land use will lead to the increase or decrease in SOM [15,16].

The government declared that a hilly region in Sri Lanka (7°07.485' N, 81°02.740' E) was degraded as a result of erosion; the reason given was excessive tobacco cultivation. A decade or more of tobacco cultivation was abandoned due to continuingly decreasing yields, leaving the soils exposed. Fallowing was not possible as the farmers looked for some alternative annual crops due to loosing income from a cash crop, leaving the lands at risk of erosion. Lack of investments on these annuals (e.g., maize and mungbean) made low profits causing a struggle for livelihood. A foundation was set up to introduce an agroforestry system based on Gliricidia (*Gliricidia sepium*) in 2001/2002 with the objective of replenishing the degraded landscape and simultaneously providing direct and indirect economic benefits. An important question arose as to the indicators of erosion with respect to soil parameters and plant productivity. Using homegardens as a benchmark for the long-term high input of green manure, it was shown that regional soils are potentially fertile for crop production, at least on flat to moderate slopes [17]. It was concluded that an assumed impact of erosion on soil fertility can only be detected in an on-farm approach, as plant and soil parameters with yield varied widely and proof of a high degree of soil erosion were scanty.

Our hypothesis was: a decrease in soil fertility due to erosion is stronger with greater inclinations of the slope and the potential loss of inherent soil chemical fertility becomes more visible, especially after nutrient replenishment through fertilizers.

## 2. RESULTS AND DISCUSSION

The top soil layer (A = ploughed horizon) was of normal thickness up to 30% inclination (Table 1).Very thin A horizons gave best proof of active erosion at inclinations above 30%. The variability of horizon thickness in all farms was very high. A thin A horizon also occurred in one flat area. This explains the low significance among the three ranges of inclination. There was no indication of former soil erosion below the A horizon, which was the same thickness in flat and steep areas. The rooting depth was shallow (<60 cm) and did not vary with inclination (data not shown). Thus, soil analyses of the upper layers (0–30 cm) were enough to explain the relationships between crop yield and soil fertility along the range of inclinations. For chemical and physical soil parameters as well as for grain yield, values for 0 to 10% inclination were set as 100%. In this range, SOM (soil organic matter), TN (total nitrogen), P (phosphorous), K (potassium) and BD (bulk density) were 12 g $kg^{-1}$, 17 mg $kg^{-1}$, 26 mg $kg^{-1}$, 19 mg $kg^{-1}$ and 1.44 g $cm^{-3}$, respectively (Table 2). An important and unexpected result was the considerable variation in each parameter at all inclinations, as revealed by the large standard deviations. Therefore, even large differences among inclination levels were usually not significant, despite the large number of farms in the study. Relative reductions of SOM by 35% occurred at inclinations of 40% and above. It was similar for K, but the variance among inclinations was even greater. For SOM and K the linear regression for both inclination ranges was significant ($R^2$s of 0.31 and 0.20, respectively) (regressions not shown). The reduction in relative amounts of P with increasing inclination was even more erratic, but it did not exceed 25% up to 60% inclination. The soil pH ranged from 5.15 to 6.83 with no significant impact of inclination (data not shown). TN changed very little with increasing inclination, whereas BD increased by 10% at 40% inclination and by 20% at 60% inclination.

**Table 1.** Soil horizons and horizon thickness (cm) of 21 soil profiles representing three slope categories.

| Soil horizons | | Slope Category (Inclination %) | | |
|---|---|---|---|---|
| | | 0–10 | 11–30 | >31 |
| | A | 28 a (53) | 24 a (38) | 6 b (45) |
| | B | 42 m (43) | 56 n (60) | 44 m (68) |
| | C | 39 x (23) | 34 x (17) | 29 x (28) |

Means followed by the same letters are not significantly different at $P < 0.05$ at each horizon *i.e.* a, b for horizon A; m, n for horizon B; x for horizon C; Numbers in the parenthesis are coefficient of variation (CV %).

**Table 2.** Organic matter (OM), total nitrogen (TN), available phosphorus (P), extractable potassium (K) and bulk density (BD) of maize fields on 92 farms, and relative changes with increasing inclination levels, with 0%–10% as the benchmark of 100%.

| Inclination (%) | SOM ($g\ kg^{-1}$) | Relative Change (%) | TN ($mg\ kg^{-1}$) | Relative Change (%) | P ($mg\ kg^{-1}$) | Relative change (%) | K ($mg\ kg^{-1}$) | Relative Change (%) | BD ($g\ cm^{-3}$) | Relative Change (%) |
|---|---|---|---|---|---|---|---|---|---|---|
| 0–10, $n$ = 61 | 18.2 a (0.59) | 100 | 12.6 a (0.35) | 100 | 25.3 a (1.60) | 100 | 18.3 a (0.87) | 100 | 1.50 a (0.015) | 100 |
| 11–20, $n$ = 27 | 18.0 a (0.74) | 99 | 11.7 ab (0.36) | 93 | 27.0 a (1.18) | 107 | 18.6 a (1.05) | 101 | 1.53 a (0.023) | 102 |
| 21–30, $n$ = 13 | 15.4 ab (1.47) | 85 | 10.1 ab (0.69) | 80 | 25.2 a (1.90) | 100 | 14.2 a (1.57) | 78 | 1.56 a (0.035) | 104 |
| 31–40, $n$ = 14 | 13.4 b (1.16) | 73 | 10.9 ab (0.69) | 87 | 23.7 a (2.11) | 94 | 13.8 a (1.30) | 75 | 1.65 a (0.044) | 110 |
| 41–50, $n$ = 7 | 12.5 b (1.46) | 68 | 11.1 ab (0.83) | 89 | 24.8 a (3.27) | 98 | 14.2 a (0.97) | 77 | 1.68 a (0.067) | 112 |
| 51–60, $n$ = 3 | 11.5 b (0.80) | 63 | 8.1 ab (2.36) | 64 | 21.8 a (2.77) | 86 | 11.5 a (3.37) | 63 | 1.71 a (0.112) | 114 |
| 61–70, $n$ = 2 | 10.5 b (1.54) | 58 | 8.0 ab (2.11) | 64 | 20.4 a (2.22) | 81 | 11.9 a (2.46) | 65 | 1.74 a (0.040) | 116 |

Means followed by the same letters are not significantly different at $P < 0.05$; Numbers in parentheses are standard errors of mean.

Harsh climatic conditions and overpopulation leading to a greater demand for food make tropical ecosystems extremely vulnerable to erosion. The ability to guarantee food security and livelihoods for the poorest in the population is threatened by a decline in soil fertility in many developing countries. How can we reveal the causes of declining soil fertility? Up to 30% inclination, an acceptably thick top soil layer A reflected a long term high intensity tilling during land preparation. Above 30% inclination, thin A horizons were the strongest proofs of former erosion events. Footprints of erosion below this horizon were not found. SOM values were below 20 g $kg^{-1}$ from 0% to 20% inclination and below 14 g $kg^{-1}$ above 30% inclination. The desired value is 26 g $kg^{-1}$ in tropical soils [18,19]. This may indicate previous and more pronounced erosion at greater inclinations. It is important to consider that organic material was incorporated into the soil only at lower inclinations, leaving soil at greater inclinations more susceptible to run-off and erosion. Nitrogen content was far below the desired minimum level of 25 mg $kg^{-1}$ [18]. Intense tillage occurs during land preparation at low inclinations to accelerate mineralization of crop residues, green manures and soil organic nitrogen [20,21]. Nitrogen is more susceptible to leaching, especially when the water table is shallow [22]. Process of denitrification leads to loss of nitrates from soil by transforming nitrates to nitrous oxide and nitrogen gases. With shallow water table, it is very likely that flat lands get water logged during rainy season. Fluctuation of redox potential due to the changes of the water table [23], which contribute to the denitrification, may explain the low nitrogen content at very low inclination, despite the demonstrated high yield potential [17]. Low nitrogen contents and low yields at steep inclinations may be due to a high degree of leaching [24,25,26] at the beginning of the cropping season when precipitation is high. The phosphorus contents in this tropical region were moderately high at all inclination levels, with the relative difference between the lowest and highest inclinations just 20%; this was true even for deeper soil layers (data not shown). Thus, phosphorous content was not a good indicator of previous erosion events but did explain in parts, the good yield response to mineral fertilizer. Low potassium contents even declined by 35% from the lowest to the highest inclination. Land preparation in this region has been characterized by slashing and burning, thus adding more potassium to the erosion-prone top soil. These difference in potassium content among inclination levels was not significant, a further proof of the tremendous variation among the many tested farms. It is, thus, not possible to extrapolate the findings of field trials at only a few locations in such a hilly environment. The soil pH was in an acceptable range. At steeper locations pH tended to be lower, possibly the result of soil erosion and the leaching of cations [8,27,28]. The narrow range of soil pH may be the result of the greater exposure of calcareous parental material at high inclinations with erosion of top soil and sedimentation of more calcium-containing soil in flat regions [29]. The mild acidic conditions may have improved the availability of phosphorus, which can explain the phosphorus content of the soil in deeper layers. Soil bulk density proved to be an even more reliable physical trait for predicting previous erosion, even though it increased from flat land to the highest inclinations by 16%, which is an indication of the loss of top soil with surface flow at steep inclinations [28].

The grain yield was assessed on 92 of a total of 119 selected farms, based on all the relevant data collected during two seasons. Season had a low impact ($p = 0.051$), despite considerable differences in rainfall, and was not considered in the model. Thus, actual and relative means of yields were compared. The yields at 0% to 10% inclination were 1.04 (ZERO), 2.73 (PK) and 3.86 (NPK) tonnes per hectare for the three levels of mineral fertilization (Table 3). These values were set at 100%. At both NPK and PK the relative yields declined by 25% at an inclination of 35% and by about 45% at an inclination of 45%. At ZERO the impact of inclination was much smaller; a generally low yield level did not decline more than 20% up to highest inclinations.

Maize yield gradually declined with increasing inclination; this was most pronounced in the ZERO treatment, thus revealing the effects of climate and moisture. In Thailand the relative yield reduction was lower at higher inclinations than in Sri Lanka [30], but the annual rainfall was also higher (2000 mm). This may indicate a general decline in soil fertility due to previous erosion, which is also coupled to the nature of the highly weathered soils of the tropical region with inherently low SOM and nutrients. Coincidently, history of tobacco cultivation, low degree of mineral fertilization and vegetation removal also substantially contributed to accelerated degradations. However, even the yields from flat fields were below the optimum that can be achieved by a long term supply of organic matter, as discussed in a simultaneous study [17]. The greater potential of homegardens was enhanced by the supply of N and K as well as by improved soil physical properties due to the application of green manure. The fertility status of the farms was heterogeneous, which led to very varied yield responses, even in the same inclination range. With respect to yield potential, the quality of the soil is largely determined by the content of SOM, which based on soil properties (e.g., texture) and management history (e.g., use of green manure, years under cultivation, fallow) [31]. The range of soil quality parameters in an inclination range clearly illustrates the heterogeneity of the farms in this region. This heterogeneity in turn influences the crop response to applied fertilizers [32].

**Table 3.** Maize yields of Nitrogen, Phosphorus and Potassium (NPK, PK) recommended mineral fertilization and without fertilizer (ZERO) mineral fertilizer treatments in baseline landscape positions, and relative change using 0%–10% as the benchmark.

| Inclination (%) | Yield ($t\ ha^{-1}$) | | | | | |
|---|---|---|---|---|---|---|
| | NPK | Relative Change (%) | PK | Relative Change (%) | ZERO | Relative Change (%) |
| 0–10 ($n = 61$) | 3.82 a (0.09) | 100 | 2.38 a (0.07) | 100 | 1.32 a (0.06) | 100 |
| 11–20 ($n = 27$) | 3.34 ab (0.13) | 87 | 2.20 ab (0.12) | 93 | 1.07 ab (0.07) | 81 |
| 21–30 ($n = 13$) | 3.21 ab (0.24) | 84 | 1.96 ab (0.15) | 82 | 0.97 ab (0.10) | 74 |
| 31–40 ($n = 14$) | 3.01 b (0.20) | 79 | 1.85 b (0.15) | 78 | 0.93 b (0.09) | 70 |
| 41–50 ($n = 7$) | 2.70 b (0.41) | 71 | 1.51 b (0.18) | 64 | 0.71 b (0.12) | 54 |

Numbers in parentheses are standard errors of mean; Means followed by the same letters are not significantly different at $P < 0.05$ for similar fertilizer treatment.

The difference between the fertilizer regimes NPK and ZERO was very large, indicating poor nutrient availability even on farms at low inclinations. Considerable differences between yields at NPK and PK (*i.e.*, 1 t $ha^{-1}$) indicate that the availability of P in this region is moderately high. Nitrogen is generally the most limiting factor in tropical crop production [2,11], seemingly nitrogen was the driving factor for yield here, too. Cultivation of tobacco and later random fallowing and cropping with inherent fluctuations of soil fertility may be a reason for the very strong impact of mineral fertilizers, especially nitrogen. This responsiveness to fertilizers was influenced by the heterogeneity of the fields and the inherent soil quality; thus the weak correlation among yields at Flat and SOM, TN, P and K with fertilizer treatments cannot be fully explained.

## 3. EXPERIMENTAL SECTION

### 3.1. Experimental Sites

In October to January in 2007/2008 and 2008/2009 (Seasons 1 and 2) there were 119 fields in both seasons spread over 25 $km^2$ in a hilly region in Sri Lanka (7°07.485' N, 81°02.740' E) (270–400 MSL). Rainfall was 1325 mm (Season 1) and 919 mm (Season 2). The absolute inclination of the 119 fields was measured [33]. A set of 92 farms was chosen because complete data sets were available, they ranged from zero to 60% inclination, clustering at around zero elevation. An initial soil analysis was carried out to determine easily accessible chemical and physical properties of the soil (pH, total nitrogen, available phosphorus, extractable potassium, organic matter, bulk density) prior to crop establishment. From each field, 4–5 soil samples were randomly collected at depths of 0–10 cm, 10–20 cm, 20–30 cm and 30–60 cm, respectively. Simultaneously, an undisturbed soil sample was collected using a core-sampler from top soil (to a depth of 20 cm) for bulk density determination. For physico-chemical characterization the following soil parameters were considered: pH was measured using a 1:2.5 soil water ratio [34], organic carbon was determined by the Walkey and Black method and total nitrogen by Kjeldahl distillation [35]; soil available phosphorus was determined by Bray 1 [36] technique; extractable potassium by ammonium acetate (NH4OAC) extraction and flame photometry; bulk density was determined on the basis of [37].

On a subset of 21 farms soil profiles were opened at the flowering stage of maize (1 × 1 m in dimension) (*i.e.*, seven profiles each for <10%, 10%–30%

and >30%). Soil horizons were classified visually, and the same soil chemical and physical properties as described above were measured.

## 3.2. Experimental Set up and Plot Management

Ruwan, an open-pollinated maize (*Zea mays* L.) variety, was sown in all fields. The fields were ploughed with a conventional plough at an effective depth of 20 cm at the start of both seasons. Maize stands were carefully established at a spacing 30 cm × 45 cm, resulting in a density of 7.4 plants $m^{-2}$. The plots were weeded by hand. All the other cultural operations followed the local practice [38].

In the central area of each field there were three principal fertilizer treatments on plots of 100 $m^{-2}$: (1) Recommended fertilizer per hectare (NPK): 69 kg nitrogen (Urea), 20 kg phosphorus (Triple Super Phosphate), 50 kg potassium (Muriate of Potash); (2) Recommended P and K fertilizer only (PK); (3) No fertilizer (ZERO). The basal amount of N, P and K was applied at planting; 46 kg N $m^{-2}$ were applied as a top dressing at 4 weeks after planting.

## 3.3. Measurements

A number of vegetative parameters were measured during the vegetative period of maize until anthesis (plant height, leaf number and SPAD). Grain yield was assessed from 10 $m^{-2}$ of each plot at maturity. Furthermore, the dry weight and dry weight percentages were determined from stover and cobs after drying in an oven at 60 °C for 48 h. Other parameters (total biomass, seed yield and harvest index) were calculated from the collected data. Grain yield reflected a more pronounced impact of mineral fertilization, and the focus was placed on this parameter only.

## 3.4. Data Analysis

Maize yield and soil parameters were tested for normality and homoscedasticity. Analysis of variance was carried out to assess the effect of inclination on yield by deriving inclination classes in 10% intervals (0%–

10%, 10%–20% *etc.*) up to 70%. According to ANOVA, significant means were separated using Tukey's HSD (Honestly Significant Difference) test. The thickness of the profile horizon was tested for significance based on ANOVA: 21 profiles were separated into three groups [0%–10% (flat), 10%–30% (moderate), >30% (steep)]. The means were separated by the Student-Newman-Keuls test. All analyses were conducted at the 5% level of significance. Regression analyses were performed to derive liner relationships between inclination and organic matter, total nitrogen, extractable potassium, available phosphorus, bulk density and maize yield. The relative decline of organic matter, total nitrogen, extractable potassium, available phosphorus and bulk density and the maize yields of NPK, PK and ZERO treatments were compared with 0%–10% (flat) as the benchmark. Statistical tools from the SAS package version 9.2 (SAS institute, 2008) was used.

## 4. CONCLUSIONS

The question was: how does one uncover the footprints of soil erosion? The study was done in a hilly region where intensive tobacco production had been carried out over many years, to the detriment of the soil. The obtained answers were not clear cut, because the heterogeneity of the soil and the yield was tremendous in all inclination ranges. This justifies the on-farm approach in this study even when the high number of farms was still not sufficient to obtain significant results. It provides a quantitative picture of the extent and variation of loss of soil fertility in these tropical conditions. Indications of previous erosion were found at high elevations, including a thin A horizon, low SOM and potassium content, high bulk densities and a low impact of mineral fertilizer. Practical consequences are clear though socioeconomically difficult to be rectified—discontinue arable farming at steep inclinations, despite temporary stable yields, intensify fertilization at lower inclinations where yield response is strong. However, these are options for future regional development!

## REFERENCES

1. Doran, J.W. Renewable agriculture and food systems (editorial material). *Renew. Agric. Food Syst.* **2007**, *22*, 80.

2. Wood, S.; Sebastian, K.; Scherr, S.J. *Pilot Analysis of Global Ecosystems: Agroecosystems*; International Food Policy Research Institute and World Resource Institute: Washington, DC, USA, 2000; pp. 1–75.

3. Craswell, E.T.; Sajapongse, A.; Howlett, D.J.B. Agroforestry in the management of sloping lands in Asia and the Pacific. *Agroforest Syst.* **1998**, *38*, 121–137.

4. Wu, J.; Hobbs, R. Key issues and research priorities in landscape ecology: An idiosyncratic synthesis. *Landsc. Ecol.* **2002**, *17*, 355–365.

5. Kaihura, F.B.S.; Kullaya, I.K.; Kilasara, M. Soil quality effects of accelerated erosion and management systems in three eco-regions of Tanzania. *Soil Till. Res.* **1999**, *53*, 59–70.

6. Francia, M.; Jos, R.; Zuazo, V.H.D.; Raya, A.M. Environmental impact from mountainous olive orchards under different soil-management systems (SE Spain). *Sci. Total Environ.* **2006**, *358*, 40–60.

7. Lal, R. Enhancing crop yields in developing countries through restoration of the soil organic carbon pool in agricultural lands. *Land Degrad. Dev.* **2006**, *17*, 197–209.

8. Sparovek, G.; Schnug, E. Temporal erosion-induced soil degradation and yield loss. *Soil Sci. Soc. Am. J.* **2001**, *65*, 1479–1486.

9. Loveland, P.; Webb, J. Is there a critical level of organic matter in the agricultural soils of temperate regions: A review.*Soil Till. Res.* **2003**, *70*, 1–18.

10. Majumder, B.; Mandal, B.; Bandyopadhyay, P.K. Organic amendments influence soil organic carbon pools and rice-wheat productivity. *Soil Sci. Soc. Am. J.* **2008**, *72*, 775–785.

11. Fageria, N.K. Green manuring in crop production. *J. Plant Nutr.* **2007**, *30*, 691–719.

12. Giardina, C.P.; Sanford, R.L.; Dockersmith, I.C. The effects of slash burning on ecosystem nutrients during the land preparation phase of shifting cultivation. *Plant Soil* **2000**, *220*, 247–260.

13. Katyal, J.C.; Rao, N.H.; Reddy, M.N. Critical aspects of organic matter management in the tropics: The example of India. *Nutr. Cycl. Agroecosys.* **2001**, *61*, 77–88.

14. Steiner, C.; Teixeira, W.G.; Lehmann, J.; Nehls, T.; de Macedo, J.L.V.; Blum, W.E.H.; Zech, W. Long term effects of manure, charcoal and mineral fertilization on crop production and fertility on a highly weathered Central Amazonian upland soil. *Plant Soil* **2007**, *291*, 275–290.

15. Sisti, C.P.J.; Sanotos, H.P.; Kohhann, R.; Alves, B.J.R.; Urquiaga, S.; Boddey, R.M. Changes in carbon and nitrogen stocks in soil under 13 years of conventional or zero tillage in southern Brazil. *Soil Till. Res.* **2004**, *76*, 39–58.

16. Macedo, M.O.; Resende, A.S.; Garcia, P.C.; Boddey, R.M.; Jantalia, C.P.; Urquiaga, S.; Campello, E.F.C.; Franco, A.A. Changes in soil C and N stock and nutrient dynamics 13 years after recovery of degraded land using leguminous nitrogen-fixing trees. *For. Ecol. Manag.* **2008**, *255*, 1516–1524.

17. Egodawatta, W.C.P.; Sangakkara, U.R.; Stamp, P. Impact of green manure and mineral fertilizer inputs on soil organic matter and crop productivity in a sloping landscape of Sri Lanka. *Field Crop Res.* **2012**, *129*, 21–27.

18. Van Holm, L.H.J. Soil Organic Matter Dynamics in Sri Lanka Soil. In *Soil Organic Matter Dynamics and the Sustainability of Tropical Agriculture*; Mulongoy, K., Merckx, R., Eds.; John Wiley & Sons Ltd: Ghent, Belgium, 1991; pp. 76–126.

19. Eilers, R.J.; Lylek, G.W.; Goh, T.B.; Mapa, R.B.; Dassanayake, A.R. Development and application of a national digital soil data base for land resource planning in Sri Lanka. *J. Soil Sci. Soc. Sri Lanka* **2003**, *15*, 45–57.

20. Sainju, U.M.; Singh, B.P. Tillage, cover crop, and kill-planting date effects on corn yield and soil nitrogen. *Agron. J.***2001**, *93*, 878–886.

21. Dinnes, D.L.; Karlen, D.L.; Jaynes, D.B. Nitrogen management strategies to reduce nitrate leaching in tile-drained Midwestern soils. *Agron. J.* **2002**, *94*, 153–171.

22. Al-Kaisi, M.; Licht, M.A. Effect of strip tillage on corn nitrogen uptake and residual soil nitrate accumulation compared with no-tillage and chisel plow. *Agron. J.* **2004**, *96*, 1164–1171.

23. Pett-Ridge, J.; Silver, W.L.; Firestone, M.K. Redox fluctuations frame microbial community impacts on N-cycling rates in a humid tropical forest soil. *Biogeochem.* **2006**, *81*, 95–110.

24. Pennock, D.J. Effects of Soil Redistribution on Soil Quality: Pedon, Landscape and Regional Scale. In *Soil Quality for Crop Production and Ecosystem Health*; Gregoritch, E.G., Carter, M.R., Eds.; Elsevier: Amsterdam, The Netherlands, 1997; pp. 167–185.

25. Cotching, W.E.; Hawkins, K.; Sparrow, L.A.; McCorkell, B.E.; Rowley, W. Crop yields and soil properties on eroded landscape of red ferrosols in north-west Tasmania. *Aust. J. Soil Res.* **2002**, *40*, 625–642.

26. Vezina, K.; Bonn, F.; Van, C.P. Agricultural land-use patterns and soil erosion vulnerability of watershed units in Vietnam's northern highlands. *Lands. Ecol.* **2006**, *21*, 1311–1325.

27. Lal, R. Deforestation and land-use effects on soil degradation and rehabilitation in western Nigeria. 2. Soil chemical properties. *Land Degrad. Dev.* **1996**, *7*, 99–119.

28. Valentin, C.; Rajot, J.L.; Mitja, D. Responses of soil crusting, runoff and erosion to fallowing in the sub-humid and semi-arid regions of West Africa. *Agric. Ecosyst. Environ.* **2004**, *104*, 287–302.

29. De Silva, G.R.R.; Dassanayaka, A.R.; Mapa, R.B. Soils of Mid Country Intermediate Zone. In *Soils of the Intermediate Zone of Sri Lanka*; Special Publication 4; Mapa, R.B., Dassanayake, A.R., Nayakekorale, H.B., Eds.; Soil Science Society of Sri Lanka: Peradeniya, Sri Lanka, 2005; pp. 105–150.

30. Sipaseuth, N.; Attanandana, T.; Vichukit, V. Subsoil nitrate and maize root distribution in two important maize soils in Thailand. *Soil Sci.* **2007**, *172*, 861–875.

31. Tittonell, P.; Vanlauwe, B.; Corbeels, M. Yield gaps, nutrient use efficiencies and response to fertilisers by maize across heterogeneous smallholder farms of western Kenya. *Plant Soil* **2008**, *313*, 19–37.

32. Zingore, S.; Murwira, H.K.; Delve, R.J. Influence of nutrient management strategies on variability of soil fertility, crop yields and nutrient balances on smallholder farms in Zimbabwe. *Agric. Ecosyst. Environ.* **2007**, *119*, 112–126.

33. Bandara, K.R.M.U.; Samarakoon, L.; Shrestha, R.P.; Kamiya, Y. Automated generation of digital terrain model using point clouds of digital surface model in forest area. *Remote Sens.* **2011**, *3*, 845–858.

34. Thomas, G.W. Soil pH and Soil Acidity. In *Methods of Soil Analysis. Part 3. Chemical Methods*; Bigham, J.M., Bartels, J.M., Eds.; American Society of Agronomy, Soil Science Society of America (ASA-SSSA): Madison, WI, USA, 1996; pp. 475–490.

35. Van Ranst, E.; Verloo, M.; Demeyer, A.; Pauwels, J.M. *Manuals for the Soil Chemistry and Fertility Laboratory*; University of Gent: Gent, Belgium, 1999; p. 243.

36. Dharmakeerthi, R.S.; Indraratne, S.P.; Kumaragamage, D. *Manuals of Soil Sampling and Analysis*; Soil Science Society of Sri Lanka: Peradeniya, Sri Lanka, 2007; pp. 57–61.

37. Bray, R.H.; Kurtz, L.T. Determination of total, organic, and available forms of phosphorus in soils. *Soil Sci.* **1954**, *59*, 39–45.

38. *Regional Fertilizer Recommendations: Agriculture Inspectors' Hand Book*; Department of Agriculture: Peradeniya, Sri Lanka, 1995; pp. 34–35.

# Chapter 6

# Soil Erosion in Britain: Updating the Record

**John Boardman** [1,2]

[1]Environmental Change Institute, School of Geography and the Environment, University of Oxford, South Parks Road, Oxford OX1 3QY, UK

[2]Department of Environmental and Geographical Science, University of Cape Town, Rondebosch 7701, South Africa

## ABSTRACT

Concern about soil erosion on arable land in Britain dates back at least 40 years. Monitoring schemes and case studies have subsequently identified the areas at risk, the rates and frequencies and the major factors responsible for erosion. Initial concern focused on impacts on the farm and therefore on food production. Latterly the emphasis has shifted to off-farm impacts particularly reservoir sedimentation, muddy flooding of properties and the ecological damage to watercourses due to nutrient enrichment, pesticides and damage to fish spawning grounds from fine-sediment inputs. The shift has therefore been to concerns about a healthy and sustainable environment which includes soils. Government agencies, the water companies and the farming industry have lagged behind scientific studies in recognising and addressing erosion problems. Attempts at mitigation are now largely driven by the need to comply with the EU Water Framework Directive whereby watercourses must reach "good status" by 2015. Future changes in land use and climate will offer further challenges in terms of effective monitoring and compliance.

## KEYWORDS

soil erosion; Britain; on-site impacts; off-site impacts; watercourses; pollution

## 1. INTRODUCTION

Soil erosion is the detachment and transport of particles or aggregates of soil. This may be due to the action of water, wind or gravity (mass movement). In the case of water, the impact of rain ("splash"), the flow of a thin sheet of water and the action of concentrated flow forming rills or gullies will contribute to erosion [1]. The effect of thin flows of water, often referred to as "interrill flow", are far less important than that of concentrated flow. Govers and Poesen [2] suggest that on average about 20% of erosion on arable land is due to interrill flow. Wind erosion is less significant in Britain than water and seems to have declined in recent years or has been less in the news than in the 1950s and 1960s [3]. Mass movements are rarely of concern in inland areas of Britain. Recent research on the continent suggests that loss of soil attached to root crops is significant in some areas, particularly with crops such as sugar beet and potatoes [4]. The movement of soil downslope due to tillage may also be significant in both historical and present time [5].

The beginnings of awareness of a soil erosion problem in Britain dates from a short paper by Evans [6]. In the paper Evans discusses erosion of peat covered uplands, wind erosion on arable land and some recent cases of water erosion on arable land. He notes the impact of erosion on crops resulting in lower yields over long periods of time; modern farming practices are the cause of the problem. Boardman charts the progress from this seminal article in the study of erosion and soil conservation in Britain [7].

The importance of the soil erosion issue has varied through time. In the 1970s Evans, working for the Soil Survey of England and Wales, was collecting erosion data for arable land [8]. There were also contributions by

Reed in the West Midlands and Morgan's work in the Silsoe area [9,10] and a more general review by Morgan [11]. The emphasis at this time was on the threat of erosion to soil quality and therefore to our ability to sustain crop yields. In the early 1980s a gradual shift in perception began with the realisation that off-farm impacts of erosion might also be significant: Morgan [11] includes several references to flooding of villages (e.g., Shepton Beauchamp, Somerset) and Stammers and Boardman discuss the flooding of property by runoff from a winter cereal field on the South Downs [12]. Throughout this period, Evans continued to emphasise the need to monitor the extent, severity and frequency of erosion, and not to neglect the uplands.

The erosion problem in Britain is also reviewed by Evans [13], Boardman and Evans [14,15] and Boardman [7]. Most data pertains to England and Wales though Boardman and Evans include Scotland [15].

The distinction is often made between "natural" and "accelerated" rates of erosion. Natural rates occur on undisturbed landscapes which in temperate regions will generally be grassland or woodland although grassland in Britain is a plagioclimax community replacing woodland and maintained by grazing. However, rates of erosion are very low on well vegetated surfaces in comparison with cleared areas especially if those are used for arable farming (Table 1).

**Table 1.** Average annual soil losses in the Silsoe area [21].

| Average annual soil losses in the Silsoe area, Bedfordshire (t/ha) | |
|---|---|
| Bare sandy loam soil | 10–45 |
| Cereals on sandy loam soil | 0.6–24 |
| Cereals on chalky soil | 0.6–21 |
| Cereals on clay soil | 0.3–0.7 |
| Grass on sandy loam soil | 0.1–3 |
| Woodland on sandy loam soil | −0.01 |

Values are for slopes of 7 to 11 degrees, except for woodland where the slope was 20 degree.

The sensitivity of landscapes to erosion is likely to increase with the continuance of arable farming and especially if more intensive approaches are used [16,17]. Soils become more erodible and more likely to crust as levels of organic matter fall. Landscapes are more sensitive if boundaries are removed to create larger fields and blocks of the same crop allow runoff to pass from field-to-field prior to the establishment of good crop cover [18]. Sensitivity to erosion is also increased by the removal of stones from soils e.g., de-stoning for potato cultivation. Conversely, as soils become stonier through long-term erosion their sensitivity declines [19]. Ultimately, the most valuable upper soil horizons are lost or mixed with poor lower horizons by ploughing, as has occurred with peat soils in parts of the Fens. Similarly, the sensitivity of upland soils to erosion is influenced by vegetation cover which is controlled by factors such as clearance, fire, and grazing levels [20].

This article will concentrate on the major issue of water erosion on arable land. Brief discussion will be provided of wind erosion and the uplands. Issues of the impact of recreation ("footpath erosion"), coastal erosion and mass movements will not be considered. Several significant developments have occurred since the last overview including the continued shift in emphasis from on-farm to off-farm impacts of erosion. Recent work emphasises the role of small scale rainfall and runoff events promoting fine sediment losses and pollution of watercourses by nutrients and pesticides.

## 2. HISTORICAL EROSION

Historical erosion in Britain is reviewed by Evans [13,22] and Bell and Boardman [23].

Accelerated erosion has been occurring in Britain since the first clearances of primeval forest 5000–6000 years ago. There were many later episodes of clearance of secondary forest. It is worth noting that it is not clearance itself (unless accompanied by burning) that results in erosion but replacement by a more vulnerable land use—especially cultivation. Bell shows that the date of deposition of colluvium in southern British dry valleys varies greatly depending on the date of the clearance episode [24]. This clearly demonstrates that erosion was not controlled by synchronous climatic events, pace Vita-Finzi [25]. Early episodes of erosion in the uplands are recorded in lake sediments and terraces e.g., in the Lake District [26], in South Wales [27] and in the Howgills [28]. The total amount of historical erosion can be estimated by the depth of truncation of soil profiles and volume of colluvium and alluvium in valley floors [13,29]. Erosion in the past has tended to take place in areas where it exists at the present time—this is due to the influence of soil type and topography. In the uplands numbers of sheep have influenced erosion rates as they do at the present time [30]. In general, erosion has been greatest at times of population pressure on the land such as in Romano-British and Medieval times [31].

Favis-Mortlock *et al*. attempt to model 7000 years of erosion on a typical South Downs hillslope [19]. Changing climate, farming practices and land use are the input parameters. Soils that are assumed to be 1m thick 7000 years ago are reduced to 13 cm at the present time. Most erosion takes place in the Bronze and Iron Ages under conditions of winter cereal farming and occasional large storms (Figure 1). The general pattern of predicted erosion accords well with archaeological records.

Historical erosion is not easy to reconstruct. Day-by-day weather cannot be known and can only be simulated from broad-scale climatic information. In particular the timing and magnitude of major storms is not known. Information on surface soil conditions as a result of cultivation with farm implements is also limited. Soils were likely to have been far less stony and therefore more erodible than their modern counterparts. Stoniness has increased as soils thinned and bedrock has been ploughed-up [19]. The specific effect of fallow (weedy or not?) and the role of field-edge tracks as sites of enhanced runoff, is not known.

## 3. CURRENT EROSION IN BRITAIN

### 3.1. Assessment of Erosion

Appropriate methods of assessment depend on objectives: what is the purpose of the assessment? National scale identification of "hotspots" of erosion will require quite different temporal and spatial scale approaches compared to assessment of a "one-off" event or the risk of erosion given certain land-use choices.

Frequently used methods of assessment of erosion are (1) experimental plots (2) sediment yield of rivers (3) field measurement of erosional forms (4) remote sensing (5) tracers (6) stratigraphy, pedology and historical documents (7) expert opinion. Details and examples are given in Boardman [32]. Evans and Brazier have compared the results from field-based monitoring and a modelling approach for the same sites in England and

Wales [33]. The model seriously over-predicts erosion both in extent and amount.

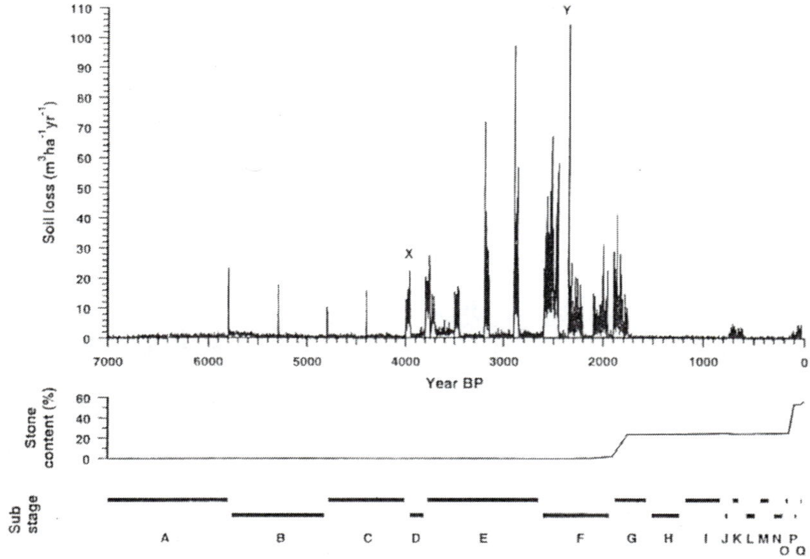


**Figure 1.** Simulated annual erosion rates for a South Downs field over the last 7000 years (Adapted from [18]).

Several reports and publications have used $^{137}Cs$ as a means of estimating erosion in England and Wales [34,35,36]. However, rates of erosion appear to be unrealistically high as compared to field measurements from the same areas. Parsons and Foster show that the assumptions on which the technique is based have not been fully tested and that validation against more traditional methods is generally lacking [37].

In the British context, assessment of erosion has been principally based on a National Monitoring Scheme (specifically for cultivated land in England and Wales) which ran from 1982 to 1986 with 17 localities selected covering most major soil associations and in total about 700 $km^2$ of farmland at risk of erosion [8,38,39,40]. The scheme was designed to assess if there was an erosion problem (which there was) but also to give information about process, cause and impacts. This is a unique scheme with most of the rest of Europe and the USA preferring model-based approaches for national assessment. The monitoring scheme for England and Wales used annual air photographs at 1:10,000 scale to identify eroding fields; these were then field checked and measurements made of rills, gullies and areas of deposition. The volume of rills and gullies can be estimated by measurement of length and cross-sectional area e.g., at 10 m intervals. Volume of fans is obtained from depth sampling and area measurement [40]. Subsequent monitoring schemes have covered much smaller areas [41,42]. A 10-year monitoring scheme on the eastern South Downs ran from 1982–1991 [43]. A very similar approach has been used for long-term soil erosion monitoring on arable land in Switzerland [44,45].

Assessment of the risk of erosion in Britain has been aided by a series of case studies of particular sites, areas or events all based on field observation and measurement:

- North Norfolk: updated with subsequent history of erosion and prevention measures at the site [46,47];
- East Shropshire: 600 sites of erosion 1967–1976 in east Shropshire [9];
- Albourne, West Sussex: erosion on strawberry crop on Lower Greensand soil [48];
- South Somerset: survey of 40 fields in winter of 1982–1983 [49];
- South Downs: surveys of erosion in two wet autumn/winters; largely on winter cereal fields [50,51];
- Oxfordshire: effects of thunderstorm on maize on sandy soil (Figure 2) [52];
- West Midlands: major summer storm on recently planted oil-seed rape fields [53];
- Eastern Scotland: erosion on arable fields [54,55];
- East Kent: erosion on loessic soils [56].

Many of these case studies deal with unusual or catastrophic erosion events and need to be placed in the context of long-term monitored distributions.

In the uplands, a series of studies by Evans highlighted the need for assessment of the impact of overgrazing which leads to ecological damage, erosion and off-site impacts on water quality [6,30,57,58]. This led to recommendations of the Royal Commission on Environmental Pollution on the need for research [59] and subsequent surveys [60,61]. In Scotland, Grieve *et al*. report on upland erosion [62,63].

## 3.2. Causes of Erosion

It is important to distinguish between ultimate or underlying causes and proximal or immediate causes. Ultimate causes are the socio-economic drivers of erosion, for example, the subsidies that encourage farmers to plant winter cereals on as much land as possible and therefore on some inappropriate land [64]. In the uplands, headage-based payments for sheep have led to overgrazing. Some erosion is driven by simple economic reality: high value crops such as potatoes, maize, strawberries may be grown in locations which in terms of erosion are highly risky but the short-term economic returns are seen as being worth the risk [65].

The immediate causes of erosion are those relating to erodibility of soils, farm management and weather events. It is all too easy to blame erosion on extreme weather events—and many of the case studies fall into this category—simply because researchers are more interested in the spectacular than in the mundane, but many cases of erosion occur as a result of relatively modest rainfall events e.g., 10 mm falling on saturated soils. Socio-economic factors explain why crops are grown in risky situations. Figure 3. shows the several factors that together combine to give erosion on winter cereals on the South Downs.

Crops are vulnerable to erosion from the time the ground is prepared to the time a sufficient crop cover exists to protect it against runoff. For different crops this time period will vary with crop growth rate and planting pattern. For example, oil-seed rape is planted close together and covers the ground very quickly: there is a low risk of erosion. In contrast, maize is planted about 0.8 m apart and takes up to three months to provide an adequate crop cover. Winter cereals on average reach a crop cover value of

30% in about two months which provides adequate protection for the soil. Unfortunately, the two month period coincides with the wettest period of the year in southern England and this coincidence is a major factor in explaining the vulnerability of winter cereals to erosion. At the same time, increases in soil moisture and cultivation-related reductions in surface roughness of the soil, all act to increase the vulnerability to runoff and erosion (Figure 4).

**Figure 2.** Faringdon, Oxfordshire: effects of May thunderstorm on maize field, see [50].

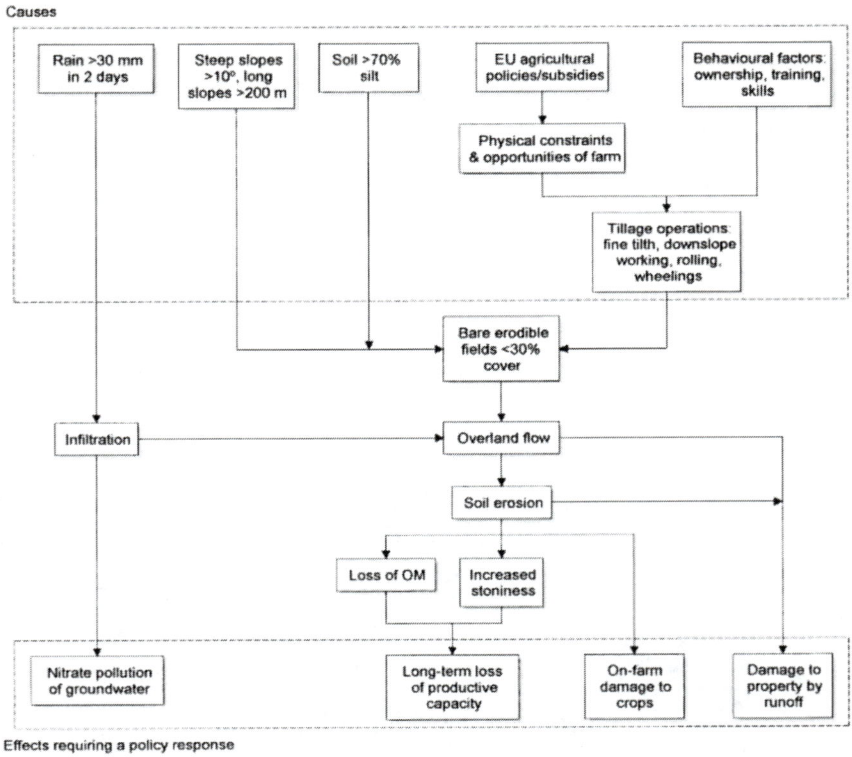


**Figure 3.** Factors influencing soil erosion on the South Downs.

The risk of erosion on various crops is listed in Evans [8]. A similar approach is advocated by Defra [66] in their advice to farmers: the combination of crops susceptible to erosion with high-risk erosion sites is to be avoided. The failure of farmers to heed this advice at a majority of sites of erosion is discussed in Boardman *et al.* [65]. Risk of erosion is also a feature of post-harvest potato and maize fields due to compaction by machinery and the tendency to leave them bare in autumn and winter months: the problem is also addressed by Defra [66]. The risk periods in relation to soil erosion for various commonly grown crops in southern England are shown in Figure 5.

For different soils and different topographic situations rainfall thresholds have to be surpassed in order to generate runoff (either due to Saturation-Excess Overland Flow (SEOF) or Infiltration-Excess Overland Flow (IEOF)—or a combination of both). In the case of the South Downs, newly planted winter cereal fields generally require about 30 mm of rainfall over a two-day period for runoff to occur. This threshold is surpassed in all years. Crusting can develop on soils after varying amounts of rainfall depending on their texture. Well crusted soils give rise to IEOF. This may occur as a result of one severe storm (e.g., 63 mm on 7 October 1987 or 100 mm on 11 October 2000 on the South Downs) or a series of rainfall events. In northern France, Papy and Boiffin suggest that around 450 mm of cumulative rainfall is required to decrease the roughness of a soil surface from "recently ploughed" to "strongly crusted"—this may take place over weeks or months [68].

Decisions made at the farm-scale greatly influence vulnerability to erosion. It is suggested that the increased use of the power-harrow to produce finer tilths [69] and declines in organic matter in soils may have contributed to erosion. Data from the national Soil Inventory of England and Wales shows carbon losses of 0.6% $yr^{-1}$ for 1978–2003 [70]. Some of the decline in organic matter may have been caused by dilution following deeper ploughing over the last 15 years [71]. These changes may have led to a decrease in aggregate stability and therefore a higher risk of erosion. A later survey confirms decreases in carbon content of arable soils in Great Britain [72]. Stark contrasts in organic levels in soils related to land use and the history of cultivation are in seen in Table 2. Compaction on fields associated with vehicle wheel tracks (wheelings) is often observed as the focus of erosion with rills developing along wheelings [73]. The impact of wheelings on runoff, sediment and nutrient losses is examined in a study by Withers et al. [74].

Loss of field boundaries and therefore increase in the size of fields and the length of slopes is undoubtedly a factor in many arable areas. But this is accompanied by the tendency for large blocks of the same crop which gives rise to bare ground at the same time of the year and therefore the possibility for flows of runoff over long distances. This is well illustrated in the case on the Sompting catchment [18] where about 60% of a 10 $km^2$ catchment was under winter cereals. Simple measures to interrupt flows with small areas of grass solved a serious off-site flooding problem.

Wind erosion is related to the enlargement of fields and the production of fine seedbeds, the increase in area of spring-planted vegetable crops and the draining and conversion to arable of former peat and sandy lands [13].

The major factor in the erosion of the uplands appears to be overgrazing largely by sheep. Evans shows that the initiation and expansion of bare soil and its subsequent erosion by the weather and animals was related to the intensity of grazing [57]. Reduction of sheep numbers reduced erosion. The high numbers of sheep are a result of the method of subsidy which led to unsustainable numbers and ecological damage to heather, and erosion [30].

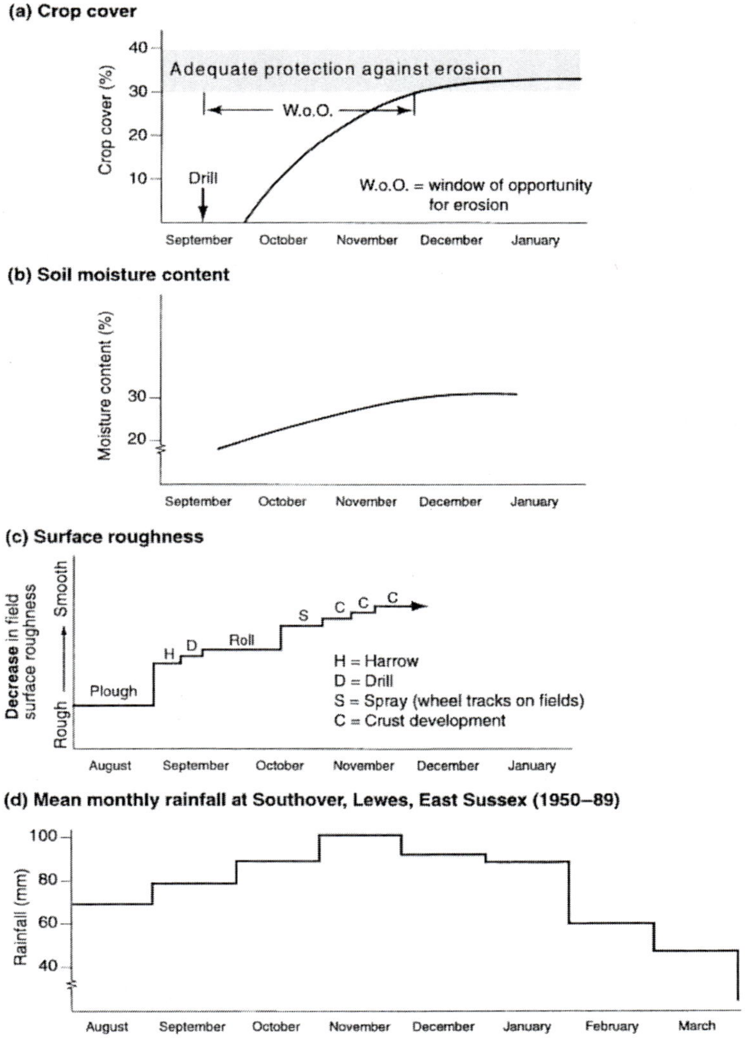


**Figure 4.** Crop cover, soil moisture, surface roughness and rainfall with reference to a “typical” winter cereal field on the eastern South Downs [43].

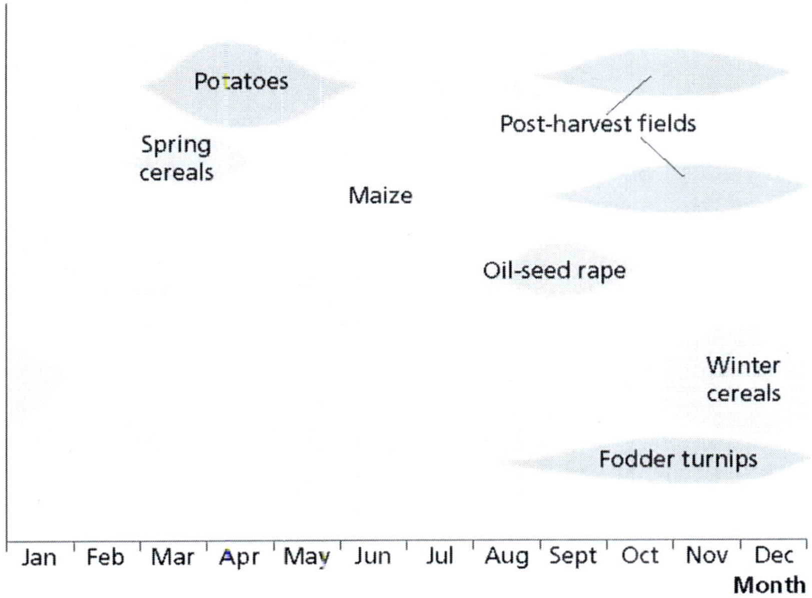


**Figure 5.** Erosion vulnerability for different types of crops. The *x*-axis length of each shape represents the time period during which soils are vulnerable to erosion. The *y*-axis height of the line denotes the risk. Highest risks are associated with periods when the shapes are thickest [67].

**Table 2.** Levels of organic carbon in South Downs soils around Houndean, Lewes, 1982–1983 [75].

| Land-use | % organic carbon (8 random samples from within each land-use block) | | | | | | | | Mean |
|---|---|---|---|---|---|---|---|---|---|
| Woodland-scrub | 6.5 | 6.4 | 6.0 | 6.2 | 6.0 | 6.5 | 6.3 | 6.3 | 6.3 |
| Permanent grassland | 12.4 | 12.0 | 11.9 | 12.2 | 11.4 | 11.8 | 11.6 | 12.2 | 11.9 |
| Converted to arable post1978 | 10.5 | 11.2 | 9.8 | 11.1 | 11.0 | 10.9 | 8.6 | 10.9 | 10.5 |
| Converted to arable 1955–1978 | 5.6 | 5.7 | 4.9 | 5.3 | 5.0 | 5.5 | 6.0 | 5.6 | 5.5 |
| Arable pre1945 | 3.4 | 2.5 | 3.2 | 3.1 | 2.5 | 3.3 | 3.1 | 3.8 | 3.1 |

## 3.3. Where is Erosion Occurring?

Several maps have been published showing areas at risk of erosion, all are based on the monitoring surveys in the 1980s and 1990s plus other published and unpublished data. All are unsatisfactory in printed form because of scale issues. Most usefully, Evans lists all 296 soil associations in England and Wales and classifies them in terms of risk [76]. For example, eight associations are "at high risk of accelerated erosion by water" and four at very high risk. The extent of the association is shown on the sheets of the National Soil Map of England and Wales at 1:250,000 scale [77] and details of the associations are in accompanying memoirs [78].

In general, certain areas are known for their susceptibility to erosion under current land use:

- Red sandstone soils of the East and West Midlands and south Devon e.g., areas around Retford (Nottinghamshire), Bridgnorth (Shropshire) and Dawlish (Devon); cases of both water and wind erosion are recorded in these areas (Figure 6).

- Chalklands of southern England: not because of the inherent erodibility of the soils but because of large areas under winter cereals, large fields and compaction along wheelings (Figure 7).

- Lower Greensand soils of Bedfordshire, West Sussex and the Isle of Wight which tend to be intensively farmed with crops such as potatoes, maize, vegetables and some cereals (Figure 8).

- Silty and fine sandy soils in Somerset e.g., on the Bridport Sands around Yeovil.

The erosion situation in Scotland has not been systematically monitored but a series of studies have highlighted a problem on arable land in the east of the country [55,69,79,80,81,82,83] and in the uplands [62,63].

Wind erosion was widespread in the Fens, sandlands of Nottinghamshire, Breckland and the Vale of York on peaty and sandy soils. Evans reports a series of major erosion events in these areas in the 1960s–1980s [3]. Such events are now less common due to changes in crops particularly the expansion of winter cereals, the loss of erodible peat soils, and the adoption of protective measures.

In terms of crop types on which erosion is occurring the data is available from the National Soil Monitoring Scheme (1982–1986) (Table 3). The majority of eroded fields are in winter cereals because of the area that the crop covers (60% of the national crop area). But rates of erosion are over twice as high in market garden crops, maize, ley grass and hops; anc significantly higher in sugar beet and potatoes.

**Figure 6.** Erosion on red sandstone soils in south Devon.

**Figure 7.** on winter cereal field, Rottingdean, East Sussex, October 1987 [50].

**Figure 8.** Erosion on Lower Greensand soils, Midhurst, West Sussex [65,67].

**Table 3.** Rates of erosion in soils drilled to different crops [1].

| Crop | No. eroded fields | Mean rate ($m^3$/ha) | Median rate ($m^3$/ha) | national crop area (%) |
|---|---|---|---|---|
| Market garden & vegetables | 102 | 5.08 | 1.47 | 3.1 |
| Maize | 26 | 4.48 | 1.00 | 0.4 |
| Ley Grass | 68 | 4.09 | 1.14 | 4.8 |
| Hops | 8 | 3.92 | 1.01 | 0.1 |
| Sugar beet | 296 | 3.04 | 0.92 | 4.4 |
| Other [2] | 49 | 2.67 | 1.07 | 1.1 |
| Potatoes | 171 | 2.53 | 1.01 | 3.2 |
| Kale & other fodder crops | 12 | 2.10 | 1.41 | 0.6 |
| Oilseed rape | 25 | 1.92 | 0.30 | 5.2 |
| Winter cereals [3] | 689 | 1.85 | 0.68 | 60.2 |
| Spring cereals [4] | 186 | 1.75 | 0.71 | 13.6 |
| Bare soil/fallow [5] | 25 | 1.61 | 0.27 | 1.1 |
| Peas | 16 | 1.21 | 0.91 | 1.3 |
| Field beans | 6 | 0.47 | 0.22 | 0.9 |

[1] Data from national soil monitoring scheme 1982–1986; [2] Crops include soft fruit, root crops for stock feed, strawberries, orchards linseed *etc*; [3] Dominantly wheat, but also barley and to a lesser extent oats and triticale; [4] Predominantly spring barley; [5] Soil surface cu ltivated but not drilled or rough fallow.

## 3.4. Rates of Erosion

Rates of erosion for eroded fields in the monitored areas are available for 1982–1986 (Table 4). Medians are preferred to means as measures of central tendency because of the skewed nature of erosion data with many more small than large events. Values tend to be low with "hotspots" in Kent, Somerset and Isle of Wight. The Kent values are probably enhanced by the use of irrigation on market garden crops. The values would have been higher had the programme extended into the wet year 1987.

A 10-year monitoring project on about 35 $km^2$ of the eastern South Downs largely on winter cereals fields, ran from 1982 to 1991 [43]. Median values vary from 0.5 to 5.0 $m^3$ $ha^{-1}$ $yr^{-1}$ (Table 5). The value of 5.0 based on 97 eroding sites contrasts with the Andover 1 value in the Table 4 because of the wetness of 1987 [50]. Very similar methods of estimating loss of soil were used in both studies. In 1987 losses of 200 $m^3$ $ha^{-1}$ were recorded on one 12 ha field at Rottingdean.

# 4. IMPACTS OF EROSION

## 4.1. On-Site Impacts

Erosion affects the operation of land managers particularly farmers in both the short and long term. Short term effects include loss of seeds and crops, inconvenience in terms of field operations especially harvesting where gullies occur, and damage due to muddy runoff to farm tracks, implements and buildings. However, rates of erosion on a particular field have probably to be in excess of 50$m^3$ $ha^{-1}$ $yr^{-1}$ and gullies to occur which inhibit operations, for impacts to be a serious consideration for the farmer. Even then, financial losses are likely to be small compared to off-site losses [84].

If we are willing to consider the longer time scale then the loss of soil will impact on our ability to grow crops. This will be especially apparent on thin soils overlying bedrock. Several studies have suggested time scales over which impacts on crop yields will become apparent [13,46,85,86]. These tend to be in the order of hundreds of years. Recent concerns for the sustainability of landscapes, and the non-agricultural services provided by soils, especially where these are not essential or ideal for arable production, suggest that we should take a broader and longer-term view of soil protection [87,88].

## 4.2. Off-Site Effects of Soil Erosion

Off-site impacts of erosion are increasingly recognised as of major concern in Europe [89] and in Britain [13,15].

Peat erosion on moorlands results in reservoir sedimentation and the costs of cleaning of dirty water before it can be used for human consumption [90]. On lowland rivers the costs of ensuring that drinking water is of adequate standard is enormous; only some of these costs are related to erosion and runoff from arable land (reviewed by Evans) [91]. There are also considerable costs related to pesticides, phosphorus and nitrate transfer from arable fields to rivers, some of which travels attached to soil particles and results in eutrophication. Cost estimates vary from £168 million [92] to £260 million per year [93].

Flooding of properties by runoff from agricultural fields ("muddy floods") has been an issue since first noted by Morgan [11] and described in some detail for one site on the South Downs by Stammers and Boardman [12]. Evans costs the impact of muddy flows on property and roads [13]. It is probably a grossly underestimated phenomenon in that few studies have attempted to collect data systematically although areas such as the Isle of Wight, the west Midlands, Somerset and Dorset, south Devon, and the South Downs regularly report flooding [94,95,96]. On the eastern South Downs 138 damaging floods have been recorded between 1976 and 2001 [43,97]. Muddy runoff from the farming of outdoor pigs is particularly prevalent in East Anglia and has resulted in a number of muddy runoff events onto roads and into houses leading to litigation [98,99].

Of growing concern has been siltation of rivers by fine sediment from arable fields. Apart from transfer of pollutants, fine sediment affects formerly gravel-bedded rivers and makes them unsuitable for fish spawning e.g., trout streams of southern England such as the western Rother, Wiltshire Avon, Test, Itchen and Allen have all reported problems [100]. Harrod reported high levels of the herbicides aldrin and dieldrin in runoff from daffodil fields reaching the Newlyn River, Cornwall, affecting eels and aquatic invertebrates [101]. This was a result of erosion during low intensity rainfall events. Evans reports on losses of nitrogen, phosphorus and atrazine to rivers and the sea [102]. Erosion on fields of potatoes, winter cereals, maize and salad crops, in the western Rother valley, West Sussex, regularly affect the river and threaten the viability of fish spawning [65,103,104].

**Table 4.** Rates of erosion at 17 monitored localities in England and Wales, 1982–1986 and of the soil associations within them with more than 30 eroded fields [15].

| Locality | Erosion ($m^3$ $ha^{-1}$) | | | Soil association | Erosion ($m^3$ $ha^{-1}$) | | | Topsoil texture |
|---|---|---|---|---|---|---|---|---|
| | Median | Mean | No | | Median | Mean | No | |
| Bedfordshire | 0.31 | 0.47 | 65 | 411d Hanslope | 0.29 | 0.58 | 49 | Clayey |
| Cumbria | 0.36 | 1.50 | 34 | - | - | - | - | - |
| Devon | 1.22 | 1.51 | 19 | - | - | - | - | - |
| Dorset | 0.81 | 1.35 | 92 | 411b Evesham 2 | 0.74 | 1.41 | 63 | Clayey |
| Gwent | 0.83 | 1.43 | 73 | 571b Bromyard | 0.80 | 1.29 | 31 | Fine silty |
| | | | | 541a Milford | 0.79 | 2.08 | 35 | Fine loamy |
| Hampshire | 1.33 | 3.95 | 59 | 571i Harwell | 1.30 | 3.52 | 55 | Loamy |
| Herefordshire | 0.68 | 1.20 | 89 | 571b Bromyard | 0.67 | 1.40 | 68 | Fine silty |
| Isle of Wight | 1.52 | 4.39 | 141 | 571g Fyfield 4 | 1.62 | 5.54 | 128 | Coarse loamy+sandy |
| Kent | 3.58 | 4.82 | 41 | - | - | - | - | - |
| Norfolk East | 0.76 | 1.03 | 118 | 551g Newport 4 | 0.50 | 1.10 | 40 | Sandy |
| | | | | 541t Wick 3 | 0.38 | 1.05 | 63 | Coarse loamy |
| Norfolk West | 0.25 | 0.85 | 110 | 343g Newmarket 2 | 0.58 | 1.12 | 60 | Coarse loamy+sandy |
| | | | | 581f Barrow | 0.19 | 0.92 | 45 | Coarse loamy |

| | | | | | | | | |
|---|---|---|---|---|---|---|---|---|
| **Nottinghamshire** | 0.71 | 1.49 | 209 | 551b Cuckney 1 | 0.72 | 2.32 | 191 | Sandy+coarse loamy |
| **Shropshire** | 0.90 | 2.36 | 197 | 572m Salwick | 1.25 | 3.33 | 36 | Fine loamy |
| | | | | 551d Newport 1 | 0.97 | 2.55 | 38 | Sandy+coarse loamy |
| | | | | 551a Bridgnorth | 0.96 | 2.76 | 94 | Sandy+coarse loamy |
| **Somerset** | 2.55 | 4.69 | 161 | 541m S. Petherton | 2.10 | 4.76 | 114 | Silty |
| | | | | 572i Curtisden | 1.39 | 2.58 | 30 | Silty |
| **Staffordshire** | 0.82 | 2.43 | 205 | 551a Bridgnorth | 1.06 | 3.15 | 100 | Sandy+coarse loamy |
| | | | | 551g Newport 4 | 0.66 | 2.60 | 53 | Sandy |
| **Sussex East** | 0.32 | 0.62 | 30 | - | - | - | - | - |
| **Sussex West** | 0.29 | 0.80 | 62 | 343h Andover 1 | 0.37 | 1.02 | 32 | Silty |

**Table 5.** Soil loss on eroding fields in the monitored area of the eastern South Downs 1982–1991 [43].

| | Median soil loss ($m^3$ $ha^{-1}$) | Total soil loss ($m^3$) | Number of sites |
|---|---|---|---|
| **82–83** | 1.7 | 1816 | 68 |
| **83–84** | 0.6 | 27 | 7 |
| **84–85** | 1.1 | 182 | 25 |
| **85–86** | 0.7 | 541 | 49 |
| **86–87** | 0.7 | 211 | 34 |
| **87–88** | 5.0 | 13529 | 97 |
| **88–89** | 0.5 | 2 | 1 |
| **89–90** | 1.4 | 940 | 51 |
| **90–91** | 2.3 | 1527 | 43 |
| **91–92** | 1.2 | 112 | 14 |

New methods of ascrib ng the source of fine-grained sediments from arable fields and river banks using fingerprinting approaches are now available and emphasise the importance of compacted areas such as wheelings [105].

Much of the current concern about fine sediment transport to watercourses associated with frequent, low magnitude runoff events is focused on unwanted nutrient enrichment and pesticide levels. Pesticide levels above present threshold levels (EU Drinking Water Directive) are frequently recorded, for example in the River Cherwell, Oxfordshire [106].

## 5. MEASURES TO PREVENT EROSION AND OFF-SITE IMPACTS

In the 1980s and early 1990s few measures were in place to control erosion. Government attitudes were that it was a localised and on-farm problem. This began to change for various reasons: crop prices declined; there was much criticism of uncontrolled farming subsidies and over-production of food crops; the growth of an effective environmental movement and an interest in the sustainable use of soils. Over-production meant that governments throughout Europe were looking to move funding for agriculture into support for environmentally-friendly approaches to landscape protection such as the Environmentally Sensitive Area scheme of the late 1980s. Thus funding, for example, was linked to reduced stocking densities with consequent conservation benefits particularly in the uplands.

The growing awareness of the off-farm impacts of erosion and externalities associated with agriculture, especially those affecting water quality [107], has led to attempts to control inputs of pollutants including sediment from agricultural fields to rivers. The European Water Framework Directive came into force in 2000 and by 2015 all inland and coastal water bodies must reach good quality or "good ecological status".

In southern England, the problem of flooding of properties by runoff from agricultural fields ("muddy floods") illustrates many of these changes without having reached an entirely satisfactory conclusion. Early measures tended to focus on the provision of emergency protection for affected properties (ditches, dams, pipes) without changing the basic cause. Principal players were local authorities who tended to negotiate on behalf of affected individuals without having a long-term interest in solutions. In some cases, houses were flooded on several occasions before effective action was taken e.g., at Sompting and Breaky Bottom on the South Downs. In some fortuitous cases, local authorities owned the farmland and could therefore put pressure on tenant farmers to change farming practices or land use in return for rent reduction e.g., Rottingdean. To some extent the problem has been lessened by a reduction in intensive arable farming in sensitive areas such as the South Downs though wet autumns such as 2000 still result in muddy flooding [43]. The focus may now have shifted to areas of continued production of high-value crops (maize, potatoes, strawberries) on erosion-prone soils e.g., the Midhurst area of West Sussex or the Lugg and Wye catchments in Herefordshire [65,67,108].

In several cases attempts have been made to sue those responsible for flooding for negligence where it must be shown that there was knowledge of the risk to the neighbour if certain practices were continued. These cases have met with varied success the issue often being the balance between unwise ("risky") land use practices and exceptional rainfall. At Breaky Bottom two out-of-court settlements persuaded the farmer responsible to change his land use including introducing trial areas of no-till cereals. In East Anglia, flooding of property by runoff from outdoor pig fields carrying soil particles and faeces resulted in a successful court case and an imposed agreement with respect to future land use [98].

At Sompting, West Sussex, a simple and relatively inexpensive change to grass on a small area of a large catchment resulted in interruption of regular down-valley floods which affected houses. In the exceptionally wet autumn of 2000 no flooding occurred at this site [18]. This seems to offer a model to many areas where muddy runoff affects watercourses and property. On the Isle of Wight, the County Council successfully used the Highways Act (1980) to persuade farmers who were generating muddy runoff from fields to take preventative action or risk being billed for protective engineering works by the council [94]. This is an approach that has been ignored by other local authorities.

The new Single Payment system of funding farmers means that they must under cross-compliance regulations fill in forms that state that action is being taken to prevent erosion and runoff. Recently revised advice from Defra lists ways in which erosion can be prevented for most crops [66].

Defra, in partnership with Natural England and the Environment Agency, has also set up the Catchment Sensitive Farming Delivery Initiative to address pollution in 50 designated catchments which suffer from water quality, pollution and erosion problems. This approach is voluntary with an emphasis on providing advice to farmers. In contrast, very effective measures are in place in parts of Flanders, Belgium, based on the provision of technical advice and grants for income-foregone and costs of mitigation measures. This approach has substantially reduced the risk of muddy flooding to villages in the area [109].

A simple approach to erosion risk assessment is proposed by Defra [66] based on the characteristics of the site: soil texture, slope and observations regarding the frequency of rills, gullies and runoff/sediment generation. The risks of particular crops on high risk sites are well documented. The approach is less good for situations where runoff and sediment passes from field to field creating off-site damage: a series of low-risk fields may give rise to serious downslope problems which this scoring system does not recognise [18,65].

Figure 5 illustrates a straightforward approach to risk management. This is based on field observation and shows the relative risk for different crops in a given area. Changes in patterns of rainfall (seasonality) will change this "average" picture. In some parts of the country because of the area under certain crops the risk will be higher on spring-planted crops e.g., maize and potatoes than on winter cereals Recent observations suggest that in areas of summer cropping the risk of runoff and erosion on post-harvest fields of maize and potatoes is being underestimated. Advice on the avoidance of such problems is clear in Defra [66] but is often ignored. In recent years maize and potatoes have often been introduced into high rainfall areas of the country which are not really suited to arable farming. The increased area under maize and outdoor pigs has probably compensated for the decline in erosion in arable cultivation since the introduction of set-aside [110].

In the uplands, erosion due to overgrazing can be controlled by reduction in sheep numbers although results may take some time before they become apparent [58,111]. Threshold numbers at which erosion occurs are discussed by Evans [58]. High numbers were encouraged by subsidies based on number (headage payments) and recent reductions in numbers have come about through area-based payments with management agreements which limit stocking e.g., through the Environmental Sensitive Area scheme.

## 6. THE FUTURE

Any assessment of current erosion is bound to be vulnerable to future changes in land use and climate [112]. Quite apart from changes related to normal rotations, changes in land use can happen rapidly driven by changes in prices and policies. Areas planted to oil-seed rape and linseed seem particularly susceptible to year-on-year change. More long term trends that clearly affect the risk of erosion are the expansion of maize in England (partly driven by warmer summers) and the growth in outdoor pig production.

Climate scenarios for the period up to 2080 suggest winter precipitation increases of up to 30% in southern England and warmer and drier summers [113]. The erosion risks would seem to be associated with more summer thunderstorms affecting spring-planted crops and a continuing threat of erosion on winter cereals. The recent trend to poly-tunnel cultivation of strawberries, and potatoes on or near flood plains in the Wye and Lugg catchments, has led to damaging runoff affecting watercourses, roads and houses. Climate change may encourage a movement of arable farming into more upland areas: conversion of grazing land to arable runs the risk of erosion in wetter years [114]. Greater emphasis is likely on the services provided by soils particularly their role in carbon storage and in flood mitigation, both of which have close links to land use and therefore the risks of soil erosion.

Evans has argued that the growing intensification of agriculture over the last 50 years and its focus on highly mechanised food production has led to significant increases in erosion. In recent years agri-environment schemes, and the actions of individual farmers have reduced erosion but these gains have been off-set by the expansion of maize and outdoor pigs. Closely associated with erosion is the extensive movement of nutrients and pesticides from fields to watercourses resulting in pollution levels that regularly exceed EU thresholds. "Either regulations or how the land is farmed will have to change" [17].

## 7. CONCLUSIONS

It is probable that there is more information available about erosion and its consequences for Britain than for any other country and most of this has been gathered by field studies rather than modelling exercises [115]. The importance of the monitoring schemes has been emphasised and the need to continue with surveillance in times of changing land use and climate is clear. Over the last thirty years the emphasis has shifted from concern about on-farm to off-farm impacts of erosion. This is likely to continue due to the emergence of water quality as an issue and the Water Framework Directive with its series of targets for improvement over the next ten years. This challenge is something that the agricultural industry and the Environment Agency will have to face. The need to address catchments as a whole in water management is not a new theme in geomorphological writing and it is relevant to the issue of soil erosion since the sources and routeways of sediment reaching watercourses need to be identified. Maybe the era of "end-of-pipe solutions" is coming to an end?

## ACKNOWLEDGEMENTS

I thank Bob Evans for many useful comments on this chapter. Figure 5 is reproduced from *Geography Review* with permission.

## REFERENCES

1. Bergsma, E. Terminology for Soil Erosion and Conservation. In Presented at International Society for Soil Science, Wageningen, The Netherlands, 1996.

2. Govers, G.; Poesen, J. Assessment of the interrill and rill contributions to total soil loss from an upland field plot.*Geomorphology* **1988**, *1*, 343–354.

3. Evans, R.; McLaren, D. *Soil Erosion and Its Impacts in England and Wales*; Friends of the Earth: London, UK, 1996; p. 31.

4. Ruysschaert, G.; Poesen, J.; Verstraeten, G.; Govers, G. Soil losses due to crop harvesting in Europe. In *Soil Erosion in Europe*; Boardman, J., Poesen, J., Eds.; Wiley: Chichester, UK, 2006; pp. 609–621.

5. Van Oost, K.; Govers, G. Tillage erosion. In *Soil Erosion in Europe*; Boardman, J., Poesen, J., Eds.; Wiley: Chichester, UK, 2006; pp. 599–608.

6. Evans, R. The need for soil conservation. *Area* **1971**, *3*, 20–23.

7. Boardman, J. The need for soil conservation in Britain—Revisited. *Area* **2002**, *34*, 419–427.

8. Evans, R. Monitoring water erosion in lowland England and Wales—A personal view of its history and outcomes.*Catena* **2005**, *64*, 142–161.

9. Reed, A.H. Accelerated erosion of arable soils in the United Kingdom by rainfall and run-off. *Outlook Agric.* **1979**, *10*, 41–48.

10. Morgan, R.P.C. Soil erosion in the United Kingdom: field studies in the Silsoe area 1973–75. In *Occasional Paper, No. 4*; National College of Agricultural Engineering: Cranfield, UK, 1977.

11. Morgan, R.P.C. Soil erosion and conservation in Britain. *Prog. Phy. Geog.* **1980**, *4*, 24–47.

12. Stammers, R.; Boardman, J. Soil erosion and flooding on downland areas. *Surveyor* **1984**, *164*, 8–11.

13. Evans, R.; McLaren, D. *Soil Erosion and Its Impacts in England and Wales*; Friends of the Earth: London, UK, 1996; p. 121.

14. Boardman, J.; Evans, R. Soil Erosion in Britain: A Review. In *Conserving Soil Resources: European Perspectives*; Rickson, R.J., Ed.; CAB International: Wallingford, UK, 1994; pp. 3–12.

15. Boardman, J.; Evans, R. Britain. In *Soil Erosion in Europe*; Boardman, J., Poesen, J., Eds.; Wiley: Chichester, UK, 2006; pp. 439–453.

16. Boardman, J. The Sensitivity of Downland Arable Land to Erosion by Water. In *Environmental Sensitivity*; Thomas, D.S.G., Allison, R.J., Eds.; Wiley: Chichester, UK, 1993; pp. 211–228.

17. Evans, R. Towards sustainable land management in Britain. *Landwards* **2013**, *Spring*, 16–17.

18. Evans, R.; Boardman, J. The curtailment of flooding in the Sompting catchment. *Soil Use Man.* **2003**, *19*, 223–231.

19. Favis-Mortlock, D.T.; Boardman, J.; Bell, M. Modelling long-term anthropogenic erosion of a loess cover, South Downs, UK. *Holocene* **1997**, *7*, 79–89.

20. Evans, R. Sensitivity of British Landscape to Erosion. In *Environmental Sensitivity*; Thomas, D.S.G., Allison, R.J., Eds.; Wiley: Chichester, UK, 1993; pp. 189–210.

21. Morgan, R.P.C. Soil erosion in Britain: The loss of a resource. *Ecologist* **1986**, *16*, 40–41.

22. Evans, R. Land Use and Accelerated Soil Erosion by Water in a Small catchment on the South Downs, West Sussex, England—Past and Present. In *Landscapes Through the Lens*; Cowley, D.C., Standring, R.A., Abicht, M.J., Eds.; Oxbow Books: Oxford, UK, 2010; pp. 129–142.

23. Bell, M.; Boardman, J. *Past and Present Soil Erosion: Archaeological and Geographical Perspectives*; Oxbow Monograph 22; Oxbow Books: Oxford, UK, 1992.

24. Bell, M. The Effects of Land-Use and Climate on Valley Sedimentation. In *Climate Change in Later Prehistory*; Harding, A.F., Ed.; Edinburgh University Press: Edinburgh, UK, 1982; pp. 127–142.

25. Vita-Finzi, C. *The Mediterranean Valleys: Geological Changes in Historical Times*; Cambridge University Press: Cambridge, UK, 1969.

26. Pennington, W. Vegetation History in the North West of England: A Regional Synthesis. In *Studies in the Vegetational History of the British Isles*; Walker, D., West, R.G., Eds.; Cambridge University Press: Cambridge, UK, 1970; pp. 41–79.

27. Jones, R.; Benson-Evans, K.; Chambers, F.M. Human influence upon sedimentation in Llangorse Lake, Wales. *Earth Surf. Proc. Land.* **1985**, *10*, 227–235.

28. Harvey, A.M.; Oldfield, F.; Baron, A.F. Dating of post-glacial landforms in the central Howgills. *Earth Surf. Proc. Land.***1981**, *6*, 401–412.

29. Evans, R. Erosion in England and Wales—The Present is the Key to the Past. In *Past and Present Soil Erosion: Archaeological and Geographical Perspectives*; Oxbow Monograph 22; Bell, M., Boardman, J., Eds.; Oxbow Books: Oxford, UK, 1992; pp. 53–66.

30. Evans, R. Soil erosion in the UK initiated by grazing animals. *Appl. Geog.* **1997**, *17*, 127–141.

31. Evans, R. Soil Erosion: Its Impact on the English and Welsh Landscape since Woodland Clearance. In *Soil Erosion on Agricultural Land*; Boardman, J., Foster, I.D.L., Eds.; Wiley: Chichester, UK, 1990; pp. 231–254.

32. Boardman, J. Soil Erosion: The Challenge of Assessing Variation through Space and Time. In *Geomorphological Variations*; Goudie, A.S., Kalvoda, J., Eds.; Nakladatelsti P3K: Prague, Czech Republic, 2007; pp. 205–220.

33. Evans, R.; Brazier, R. Evaluation of modelled spatially distributed predictions of soil erosion by water versus field-based assessments. *Environ. Sci. Pol.* **2005**, *8*, 493–501.

34. Quine, T.A.; Walling, D.E. Rates of soil erosion on arable fields in Britain: Quantitative data from caesium-137 measurements. *Soil Use Man.* **1991**, *7*, 169–176.

35. Walling, D.E.; Quine, T.A. Use of $^{137}Cs$ measurements to investigate soil erosion on arable fields in the UK: Potential applications and limitations. *J. Soil Sci.* **1991**, *42*, 147–165.

36. Quine, T.A.; Walling, D.E. *In Farm Land Erosion: In Temperate Plains Environment and Hills*; Wicherek, S., Ed.; Elsevier: Amsterdam, The Netherlands, 1993; pp. 357–371.

37. Parsons, A.J.; Foster, I.D.L. What can we learn about soil erosion from the use of $^{137}Cs$? *Earth Sci. Rev.* **2011**, *108*, 101–113.

38. Evans, R. Assessing erosion in England and Wales. In Proceedings of the 7th ISCO ConferenceInternational Soil Conservation Organisation, Sydney, Australia, 27–30 September 1992; pp. 82–91.

39. Evans, R. Extent, Frequency and Rates of Rilling of Arable Land in Localities in England and Wales. In *Temperate Plains Environments and Hills*; Wicherek, S., Ed.; Elsevier: Amsterdam, The Netherlands, 1993; pp. 177–190.

40. Evans, R.; Boardman, J. Assessment of Water Erosion in Farmers' Fields in the UK. In *Conserving Soil Resources: European Perspectives*; Rickson, R.J., Ed.; CABI: Wallingford, UK, 1994; pp. 13–24.

41. Chambers, B.J.; Davies, D.B.; Holmes, S. Monitoring of water erosion on arable farms in England and Wales, 1989–90.*Soil Use Man.* **1992**, *8*, 163–170.

42. Chambers, B.J.; Garwood, T.W.D. Monitoring of water erosion on arable farms in England and Wales, 1990–94. *Soil Use Man.* **2000**, *16*, 93–99.

43. Boardman, J. Soil erosion and flooding on the eastern South Downs, southern England, 1976–2001. *Trans. Inst. Brit. Geog. NS* **2003**, *28*, 176–196.

44. Prasuhn, V. Soil erosion in the Swiss midlands: Results of a 10 year field survey. *Geomorphology* **2011**, *126*, 32–41.

45. Prasuhn, V. On-farm effects of tillage and crops on soil erosion measured over 10 years in Switzerland. *Soil Till. Res.***2012**, *120*, 137–146.

46. Evans, R.; Nortcliff, S. Soil erosion in north Norfolk. *J. Agric. Sci. Camb.* **1978**, *90*, 185–192.

47. Evans, R. Curtailing water erosion of cultivated land: An example from north Norfolk, eastern England. *Earth Surf. Proc. Land.* **2006**, *31*, 598–605.

48. Boardman, J. Soil erosion at Albourne, West Sussex, England. *Appl. Geog.* **1983**, *3*, 317–329.

49. Colborne, G.J.N.; Staines, S.J. Soil erosion in south Somerset. *J. Agric. Sci. Camb.* **1985**, *104*, 107–112.

50. Boardman, J. Severe erosion on agricultural land in East Sussex, UK October 1987. *Soil Tech.* **1988**, *1*, 333–348.

51. Boardman, J. Storms, floods and soil erosion on the South Downs, East Sussex, autumn and winter 2000–01. *Geography***2001**, *84*, 346–355.

52. Boardman, J.; Burt, T.; Evans, R.; Slattery, M.C.; Shuttleworth, H. Soil erosion and flooding as a result of a summer thunderstorm in Oxfordshire and Berkshire, May 1993. *Appl. Geog.* **1996**, *16*, 21–34.

53. Foster, I.D.L.; Harrison, S.; Clark, D. Soil erosion in the West Midlands. *Geography* **1997**, *82*, 231–239.

54. Watson, A.; Evans, R. A comparison of estimates of soil erosion made in the field and from photographs. *Soil Till. Res.***1991**, *19*, 17–27.

55. Watson, A.; Evans, R. Water erosion of arable fields in north-east Scotland, 1985–2007. *Scott. Geog. J.* **2007**, *123*, 107–121.

56. Boardman, J.; Hazelden, J. Examples of erosion on brickearth soils in east Kent. *Soil Use Man.* **1986**, *2*, 105–108.

57. Evans, R. Overgrazing and soil erosion on hill pastures with particular reference to the Peak District. *J. Brit. Grass. Soc.***1977**, *32*, 65–76.

58. Evans, R. The erosional impacts of grazing animals. *Prog. Phy. Geog.* **1998**, *22*, 251–268.

59. Royal Commission on Environmental Pollution, *Sustainable Use of Soil*; Report No 19; Her Majesty's Stationery Office: London, UK, 1996.

60. McHugh, M.; Harrod, T.; Morgan, R. The extent of soil erosion in upland England and Wales. *Earth Surf. Proc. Land.***2002**, *27*, 99–107.

61. McHugh, M. Short-term changes in upland soil erosion in England and Wales: 1999 to 2002. *Geomorphology* **2007**, *86*, 204–213.

62. Grieve, I.C.; Hipkin, J.A.; Davidson, D.A. *Soil Erosion Sensitivity in Upland Scotland*; Report No 24; Scottish Natural Heritage Research Survey and Monitoring Report: Edinburgh UK, 1994.

63. Grieve, I.C.; Davidson, D.A.; Gordon, J.E. Nature, extent and severity of soil erosion in upland Scotland. *Land Deg. Rehab.* **1995**, *6*, 41–55.

64. Boardman, J.; Poesen, J.; Evans, R. Socio-economic factors in soil erosion and conservation. *Environ. Sci. Pol.* **2003**, *6*, 1–6.

65. Boardman, J.; Shepheard, M.; Walker, E.; Foster, I.D.L. Soil erosion and risk assessment for on- and off-farm impacts: A test case in the Midhurst area, West Sussex, UK. *J. Environ. Man.* **2009**, *90*, 2578–2588.

66. *Controlling Soil Erosion*; Department for Environment, Food and Rural Affairs: London, UK, 2005.

67. Boardman, J. Soil erosion and farming: Assessing the risk. *Geog. Rev.* **2012**, *26*, 36–40.

68. Papy, F.; Boiffin, J. The use of farming systems for the control of runoff and erosion. *Soil Technol.* **1989**, *1*, 29–38.

69. Speirs, R.B.; Frost, C.A. The increasing incidence of accelerated soil water erosion on arable land in the east of Scotland. *Res. Dev. Agric.* **1985**, *2*, 161–167.

70. Bellamy, P.H.; Loveleand, P.J.; Bradley, R.I.; Lark, R.M.; Kirk, G.J. Carbon losses from all soils across England and Wales 1978–2003. *Nature* **2005**, *437*, 245–248.

71. MAFF, Resource Use—Organic Matter Content of Agricultural Topsoils. In *Towards Sustainable Agriculture: A Pilot Set of Indicators*; Ministry of Agriculture, Fisheries and Food: London, UK, 2000; Volume PB4583, pp. 47–48.

72. Emmett, B.A.; Reynolds, B.; Chamberlain, P.M.; Rowe, E.; Spurgeon, D.; Brittain, S.A.; Frogbrook, Z.; Hughes, S.; Lawlor, A.J.; Poskitt, J.; *et al.* *Countryside Survey: Soils Report from 2007*; Technical Report No. 9/07; Centre for Ecology & Hydrology: Wallingford, UK, 2010.

73. Reed, A.H. The erosion risk of compaction. *Soil Wat.* **1983**, *1*, 29–33.

74. Withers, P.J.A.; Hodgkinson, R.A.; Bates, A.; Withers, C.M. Some effects of tramlines on surface runoff, sediment and phosphorus mobilization on an erosion-prone soil. *Soil Use Man.* **2006**, *22*, 245–255.

75. Boardman, J.; Robinson, D.A. Soil erosion, climatic vagary and agricultural change on the Downs around Lewes and Brighton, autumn 1982. *Appl. Geog.* **1985**, *5*, 243–258.

76. Evans, R. Soils at risk of accelerated erosion in England and Wales. *Soil Use Man.* **1990**, *6*, 125–131.

77. Soil Survey of England and Wales, *National Soil Map*; Soil Survey of England and Wales: Harpenden, UK, 1983.

78. Jarvis, M.G.; Allen, R.H.; Fordham, S.J.; Hazelden, J.; Moffat, A.J.; Sturdy, R.G. *Soils and Their Use in South East England*; Bulletin No. 15; Soil Survey of England and Wales: Harpenden, UK, 1984.

79. Frost, C.A.; Speirs, R.B. Water erosion of soils in south-east Scotland—A case study. *Res. Dev. Agric.* **1984**, *1*, 145–152.

80. Frost, C.A.; Speirs, R.B. Soil erosion from a single rainstorm over an area in East Lothian, Scotland. *Soil Use Man.* **1996**,*12*, 8–12.

81. Speirs, R.B.; Frost, C.A. Soil water erosion on arable land in the United Kingdom. *Res. Dev. Agric.* **1987**, *4*, 1–11.

82. Kirkbride, M.P.; Reeves, A.D. Soil erosion caused by low-intensity rainfall in Angus, Scotland. *Appl. Geog.* **1993**, *13*, 299–311.

83. Davidson, D.A.; Harrison, D.J. The nature, causes and implications of water erosion on arable land in Scotland. *Soil Use Man.* **1995**, *11*, 63–68.

84. Robinson, D.A.; Blackman, J.D. Some Costs and Consequences of Soil Erosion and Flooding around Brighton and Hove, Autumn 1987. In *Soil Erosion on Agricultural Land*; Boardman, J., Foster, I.D.L., Dearing, J.A., Eds.; Wiley: Chichester, UK, 1990; pp. 369–382.

85. Evans, R. Assessments of Soil Erosion and Peat Wastage for Parts of East Anglia, England, a Field Visit. In *Soil Conservation: Problems and*

*Prospects*; Morgan, R.P.C., Ed.; Wiley: Chichester, UK, 1981; pp. 521–530.

86. Morgan, R.P.C. Sensitivity of European Soils to Ultimate Physical Degradation. In *Scientific Basic for Soil Protection in the European Community*; Barth, H., L'Hermite, P., Eds.; Elsevier: London, UK, 1987; pp. 147–157.

87. Evans, R. Sustainable practices to limit soil erosion: A review and discussion. *CAB Rev. Perspect. Agric. Vet. Sci. Nutr. Nat. Resour.* **2006**, *1*, 030:1–030:12.

88. Boardman, J. Soils and Sustainability: The Future of the South Downs. In *Sustainable Farmland Management: Transdisciplinary Approaches*; Fish, R., Seymour, S., Watkins, C., Steven, M., Eds.; CAB International: Wallingford, UK, 2008; pp. 49–57.

89. Boardman, J.; Poesen, J. Soil Erosion in Europe: Major Processes, Causes and Consequences. In *Soil Erosion in Europe*; Boardman, J., Poesen, J., Eds.; Wiley: Chichester, UK, 2006; pp. 479–487.

90. Butcher, D.P.; Labadz, J.C.; Potter, W.R.; White, P. Reservoir sedimentation rates in the Southern Pennine region, UK. In *Geomorphology and Sedimentology of Lakes and Reservoirs*; McManus, J., Duck, R.W., Eds.; Wiley: Chichester, UK, 1993; pp. 73–92.

91. Evans, R.; McLaren, D. *Soil Erosion and Its Impacts in England and Wales*; Friends of the Earth: London, UK, 1996; p. 52.

92. Environment Agency, *Agriculture and Natural Resources: Benefits, Costs and Potential Solutions*; Environment Agency: Bristol, UK, 2002.

93. Evans, R. Soil Erosion and Land use: Towards a Sustainable Policy. In Proceedings of the Professional Environmental SeminarCambridge Committee for Interdisciplinary Environmental Studies, Cambridge, UK, 1995; Evans, R., Ed.; White Horse Press: Cambridge, UK, 1995; pp. 14–26.

94. Boardman, J. Property damage by run-off from agricultural land. *Town Coun. Plan.* **1994**, *63*, 249–251.

95. Boardman, J. The problem of muddy floods. *Rural Prop. Bull.* **2000**, *November/December*, 26–27.

96. Boardman, J. Flooding and the use of land. *Town Coun. Plan.* **2001**, *70*, 113.

97. Boardman, J.; Evans, R.; Ford, J. Muddy floods on the South Downs, southern England: Problem and response.*Environ. Sci. Pol.* **2003**, *6*, 1–6.

98. Evans, R. Outdoor pigs and flooding: An English case study. *Soil Use Man.* **2004**, *20*, 178–181.

99. Evans, R. Runoff and soil erosion in arable Britain: Changes in perception and policy since 1945. *Environ. Sci. Pol.* **2010**,*13*, 141–149.

100. Evans, R.; McLaren, D. *Soil Erosion and its Impacts in England and Wales*; Friends of the Earth: London, UK, 1996; p. 53.

101. Harrod, T.R. Runoff, Soil Erosion and Pesticide Pollution in Cornwall. In *Conserving Soil Resources: European Perspectives*; Rickson, R.J., Ed.; CAB International: Wallingford, UK, 1994; pp. 105–115.

102. Evans, R. Rural land use in England and Wales and the delivery to the adjacent seas of nitrogen, phosphorus and atrazine. *Soil Use Man.* **2002**, *19*, 1–7.

103. Theurer, F.D.; Harrod, T.R.; Theurer, M. *Sedimentation and Salmonids in England and Wales*; R&D technical report P194; Environment Agency: Rotherham, UK, 1998; p. 63.

104. Sear, D.A. *Fine Sediment Accumulations in the River Rother, West Sussex*; Report to Southern Region NRA: Southampton, UK, 1996; p. 27.

105. Collins, A.L.; Zhang, Y.S.; Duethmann, D.; Walling, D.E.; Black, K.S. Using a novel tracing-tracking framework to source fine grained sediment loss to watercourses at sub-catchment scale. *Hydrol. Proc.* **2013**, *27*, 959–974.

106. Evans, R. Pesticide run off into English rivers—A big problem for farmers. *Pesticide News* **2009**, *85*, 12–15.

107. Pretty, J.N.; Brett, C.; Gee, D.; Hine, R.E.; Mason, C.F.; Morison, J.I.L.; Raven, H.; Rayment, M.D.; van der Bijl, G. An assessment of the total external costs of UK agriculture. *Agric. Syst.* **2000**, *65*, 113–136.

108. Walker, E. Soil Erosion and DEFRA's Catchment Sensitive Farming Programme—A Hertfordshire Case Study. Master Thesis, Environmental Change Institute, University of Oxford, UK, 2007.

109. Boardman, J.; Vandaele. K. Soil erosion, muddy floods and the need for institutional memory. *Area* **2010**, *42*, 502–513.

110. Evans, R. Land use, Sediment Delivery and Sediment Yield in England and Wales. In *Soil Erosion and Sediment Redistribution in River Catchments*; Owens, P.N., Collins, A.J., Eds.; CAB International: Wallingford, UK, 2006; pp. 70–84.

111. Evans, R. Curtailing grazing-induced erosion in a small catchment and its environs, the Peak District, Central Englan.*Appl. Geog.* **2005**, *25*, 81–95.

112. Macleod, C.J.A.; Falloon, P.D.; Evans, R.; Haygarth, P.M. The effects of climate change on the mobilization of diffuse substances from agricultural systems. *Adv. Agron.* **2012**, *115*, 41–77.

113. Hulme, M.; Jenkins, G.J.; Lu, X.; Turnpenny, J.; Mitchell, T.D.; Jones, R.G.; Lowe, J.; Murphy, J.M.; Hassell, D.; Boorman, P.; McDonald, R.; Hill, S. *Climate Change Scenarios for the United Kingdom: The UKCIPPO2 Scientific Report*; Scientific Report for Tyndall Centre for Climate Change Research, School of Environmental Sciences, University of East Anglia: Norwich, UK, 2002.

114. Boardman, J.; Spivey, D. Flooding and erosion in west Derbyshire, April 1983. *East Mid. Geog.* **1987**, *10*, 36–44.

115. Evans, R. Assessment and monitoring of accelerated water erosion of cultivated land—When will reality be acknowledged? *Soil Use Man.* **2013**, *24*, 105–118.

# Chapter 7

# Olive Cultivation, its Impact on Soil Erosion and its Progression into Yield Impacts in Southern Spain in the Past as a Key to a Future of Increasing Climate Uncertainty

**José A. Gómez [1,*], Juan Infante-Amate [2], Manuel González de Molina [2], Tom Vanwalleghem [3], Encarnación V. Taguas [4] and Ignacio Lorite [5]**

[1]Institute for Sustainable Agriculture, CSIC, Alameda del Obispo s/n. Córdoba 14004, Spain
[2]Agroecosystems History Laboratory, Pablo de Olavide University, Carretera de Utrera, km 1, Sevilla 41013, Spain
[3]Department of Agronomy, University of Cordoba, Campus de Rabanales, Córdoba 14014, Spain
[4]Department of Rural Engineering, University of Cordoba, Campus Rabanales, Córdoba 14014, Spain
[5]IFAPA-Alameda del Obispo. Avda. Menenez Pidal s/n. Córdoba 14004, Spain

## ABSTRACT

This article is intended as a review of the current situation regarding the impact of olive cultivation in Southern Spain (Andalusia) on soil degradation processes and its progression into yield impacts, due to diminishing soil profile depth and climate change in the sloping areas where it is usually cultivated. Finally, it explores the possible implications in the regional agricultural policy these results might have. It tries to show how the expansion and intensification of olive cultivation in Andalusia, especially since the late 18th century, had as a consequence an acceleration of erosion processes that can be identified by several indicators and techniques. Experimental and model analysis indicates that the rate of soil erosion accelerated since the expansion of mechanization in the late 1950s. In addition, that unsustainable erosion rates have prevailed in the region since the shift to a more intense olive cultivation systems by the end of the 17th Century. Although agroenvironmental measures implemented since the early 2000s have reduced erosion rates, they are still unsustainably high in a large fraction of the olive area in the region. In the case of olive orchards located in steeper areas with soils of lower water-holding capacity (due to coarse texture and stone content), cumulative erosion has already had a high impact on reducing their potential productivity. This is one of the factors that contributes towards increasing the gap between these less intensified orchards in the mountainous areas and those in the hilly areas with more gentle slopes, such as for instance the lower stretches of the Guadalquivir River Valley. In the case of olive orchards in the hilly areas with better soils, easier access to irrigation and lower production costs per unit, the efforts on soil conservation should be oriented towards limiting off-site damage, since the soil water-storage function of these soils may be preserved in the medium term even at the current soil erosion rates. The assessment made in this manuscript should be regarded as an initial approximation, since additional efforts in terms of increasing experimental records (for current or historical erosion) and of improving model analysis, with more comprehensive studies and more robust calibration and validation processes, are required.

## KEYWORDS

olive; history; soil erosion; water balance; Southern Spain; yield; climate change

## 1. INTRODUCTION

This article presents a review on the impact of olive cultivation in Southern Spain (Andalusia)—where it is the dominant crop—and on its impact on soil degradation and potential productivity in the sloping landscape areas where olives are usually cultivated. Only the impact on yield due to the reduction of soil water storage induced by of decreasing soil depth by erosion will be studied, without considering the impact of loss of soil fertility. Soil erosion in olive orchards in the region has been mentioned as one of the major threats to the sustainability of this crop [1]. The on-site damage attributed to water erosion in olive orchards in Andalusia entails a decrease in productivity in olive yield as well as added difficulties in the access to and the movement within the orchards due to gully erosion. The major off-site effects of water erosion involves the contamination by agrochemicals (especially herbicides and fertilizers) of water bodies. This has resulted in the eutrophication of reservoirs, damages to infrastructures such as siltation of dams and rivers, and an increase in flood damages due to the large amount of sediment carried away by the flows. This article attempts to provide a comprehensive review of most of the available experimental information. This information is complemented by model analyses intended to help interpreting the progression of the decreasing depth of the soil and its impact on yields, and ends up with a discussion about the future situation as regards the potential yield under projected climate change scenarios, if the current erosion rates remain unchecked. It is organized in six sections, with introduction been Section 1. Section 2 presents a brief historical introduction to the expansion of olive cultivation in the region, especially since the 18th century. Section 3 contains an evaluation of the historical evolution of water and tillage erosion rates based on a few study cases. Section 4 shows an appraisal of the current situation (regarding soil management and water erosion risk). Section 5 describes a water balance model used as a tool to evaluate the effect of soil erosion on declining soil water storage capacity in olive orchards as well as its impact on olive yield, which is in turn used in Section 6 to allow for an interpretation of the impact on yield of the cumulative soil loss since the start of olive cultivation in a few selected cases. Finally, Section 6 also explores the implications of declining soil depth in olive orchards under future climate change scenarios for Andalusia.

## 2. HISTORICAL EVOLUTION OF OLIVE CULTIVATION IN ANDALUSIA

Andalusia, as a region, has one the oldest olive-growing traditions anywhere in the world. Today, it is home to almost 1.5 Mha of olive groves, representing over 15% of the total global olive land. This is not new, between the 1st and 3rd Centuries AD, the region of Baetica (today's Andalusia) provided most of Rome's supply of olive oil [2]. Despite this long tradition, the major expansion of olive-growing in Southern Spain began in earnest in the 19th century, ushering in a wave of growth that still continues to this day. Since the mid-19th century, it has been documented that land with olive groves has increased from 0.411 Mha to the present-day figure of 1.496 Mha (Table 1). A major factor in this expansion of olive cultivation in Spain during the19th century is the fall of the so called *Ancien Régime* and the consequent implementation of liberal policies. These policies deregulated the market factors and the control by landlords who, up until then, had exercised their feudal privileges and had limited the expansion of olive-growing [3,4,5]. In addition to these institutional aspects, from the second half of the 19th century onwards, industrialization played a key role in the transformation of the olive sector. The main innovations revolved around the development of new forms of transportation, which along with the wave of globalization of global economy resulted in a favorable institutional framework for Andalusian growers, who benefitted from the growing global demand for olive oil. Between 1896 and 1930, the total land area dedicated to olive-growing in Andalusia increased by 50% (from 0.668 Mha to 0.983 Mha), production increased 2.5 fold (from 87 kt to 219 kt) and exports throughout the country expanded by 380% (from 29 kt to 86 kt), see Table 1. This period, during which Spain became the world's leader in olive oil production, surpassing Italy, is known as the "Golden Age" of the olive grove.

This period ended with the Civil War (1936/1939), which was followed by Franco's dictatorship. Both episodes ushered in a period of depression in the sector, characterized by a lack of agricultural development, the growth of the black market in all the sectors of the economy, and a widespread decline in exports [6,7]. The lack of any incentives to expand olive production led to the stagnation of the sector, a process that became even more marked in the 1960s. By this date, liberal economic policies were reinstated, but Spanish olive production found itself in a very precarious condition to compete in the global olive oil markets [8].

This downturn was followed by a new period of growth from 1970 onwards (a second "Golden Age"), when growing global demand opened up new opportunities for Spanish olive-growers. Admission to the European Union brought among other things with it, the oil production subsidy program set up as part of the Common Agricultural Policy. This factor, along with a rise in international demand, the formation of lobbies promoting olive oil and the publication of new studies linking its consumption to health benefits, led to a new period of expansion [9,10,11,12]. This was also a period when the intensification of agriculture permitted unprecedented growth in productivity. Spain went from producing 333 kt and exporting 96 kt in 1966/1970 to producing 1222 kt and exporting 643 kt in 2006/2010. Although the total land area occupied increased from 1.71 to 2.36 Mha in this period, growth in production and exports was significantly higher [4,5,11,13].

**Table 1.** Land area occupied by olive groves and oil production in Andalusia, Spain, and the World. Source: [13].

| | Total Land Area (Mha) | | | Olive Oil Production (kt) | | | Percentage of Total World Production | |
|---|---|---|---|---|---|---|---|---|
| | Andalusia | Spain | World | Andalusia | Spain | World | Andalusia | Spain |
| 1851/1860 | 0.411 | 0.857 | - | - | - | - | - | - |
| 1861/1870 | - | - | - | - | 174 | - | - | - |
| 1871/1880 | 0.657 | - | - | - | 194 | - | - | - |
| 1881/1890 | - | - | - | - | 202 | - | - | - |
| 1891/1900 | 0.616 | 1.146 | - | 115 | 208 | - | - | - |
| 1901/1910 | 0.786 | 1.349 | - | 113 | 199 | - | - | - |
| 1911/1920 | 0.823 | 1.498 | - | 172 | 283 | 684 | 25.09 | 41.34 |
| 1921/1930 | 0.926 | 1.708 | - | 213 | 339 | 812 | 26.27 | 41.74 |
| 1931/1940 | 0.986 | 1.910 | 5.500 | 202 | 320 | 853 | 23.65 | 37.46 |
| 1941/1950 | 1.050 | 1.987 | - | 219 | 314 | 948 | 23.14 | 33.08 |
| 1951/1960 | 1.108 | 2.108 | - | 259 | 374 | 1080 | 23.97 | 34.60 |
| 1961/1970 | 1.170 | 1.945 | - | 265 | 341 | 1312 | 20.18 | 25.96 |
| 1971/1980 | 1.192 | 2.090 | - | 314 | 413 | 1603 | 19.59 | 25.79 |
| 1981/1990 | 1.197 | 2.052 | 6.747 | 405 | 516 | 1681 | 24.12 | 30.72 |
| 1991/2000 | 1.301 | 2.136 | 7.800 | 565 | 720 | 2201 | 25.66 | 32.71 |
| 2001/2010 | 1.496 | 2.361 | 9.296 | 953 | 1164 | 2762 | 34.52 | 42.15 |

It should be noted that this growth concentrated around very specific geographical areas. Olive-growing has historically been centered around the Mediterranean and southern areas of the Peninsula, particularly Andalusia, and within this region, in the provinces of Jaen, Cordoba, and Seville, which have accounted for close to 50% of the total olive production area in Spain from the late 19th century to the present day (Figure 1). Between 1890 and 2010, when Spanish olive groves increased their total land area from 1.082 Mha to 2.36 Mha, 44% of the new plantations have been set up in Jaen, Cordoba, and Seville, and 66% are located in Andalusia. Andalusia currently accounts for 62% of the total olive land area in Spain and 15% in the world, making it home to the largest concentration of tree crops in Europe (Table 1).

Behind this history of expansion lie changes in productive functionality, management approaches and crop ecology. In the pre-industrial age, the morphology of the olive orchard in Southern Spain was completely different to how it is today. Recent studies have documented that between the 18th and 19th centuries, olive orchards were viewed more in terms of harnessing forest resources than as an agricultural crop [11]. They were characterized by having a small number of trees per hectare, scarce labor, little input, and therefore relatively low olive production. Hence, the role played by traditional olive production was not so much that of a commercial crop linked to the exportation of oil, but rather a harnessing of resources capable of supplying a higher number of goods for domestic purposes. Oil was not only used for human consumption but also for lighting, as an industrial lubricant, or to make soap. The wet solid waste from the oil milling process, known as “orujo”, was used for fertilization purposes, as fuel, or as animal feed. Olive wood was one of the main fuels used in olive growing areas. Furthermore, since tillage was practiced sparingly, plant cover between the trees grew in abundance and was used as animal feed, as were the leaves from the tree. In 1891/1895, of the total production used from each hectare of olive groves in Andalusia, measured in energy terms, only 11.9% was used for human consumption, whereas 18.4% was used for animal feed, and 59.0% was used as fuel (Figure 2).

This process of transition in production took place not only through a change in the demand for products, but also in the way olive groves were managed over the course of the century. There is no time-series data on a national level regarding the use of fertilizers in olive groves, though there is scattered evidence. Agronomical reports for 1888 and 1921 indicate that Spanish olive groves received practically no input in terms of fertilization beyond the nitrogen fixation offered by spontaneous leguminous plants cover and the return of pruning waste to the soil [14,15,16,17]. From the 1960s onwards, Spanish agriculture embarked on a process of intensification, with a rapid increase in the use of both inorganic fertilizers—sometimes in excess [18]—and industrial machinery. Nevertheless, there are indications that this trend towards increased fertilizer input started in some orchards in the region prior to 1960 [15]. This trend towards intensification can be noted even in the pre-industrial age: for instance, in the case of soil tillage Moldboard ploughs were not used in Andalusia until the late 19th century. Prior to that, the Roman plough was used once or twice a year. Intensification came about gradually over the course of the 19th century, and in the early 20th century, the lanes between the trees were ploughed three or four times, eliminating almost all of the plant cover. In the second half of the 20th century, industrial machinery was introduced, and in the late 20th century herbicide treatments, alone or in combination with ploughing, eliminated practically all proliferation of plant cover between the tree lines.

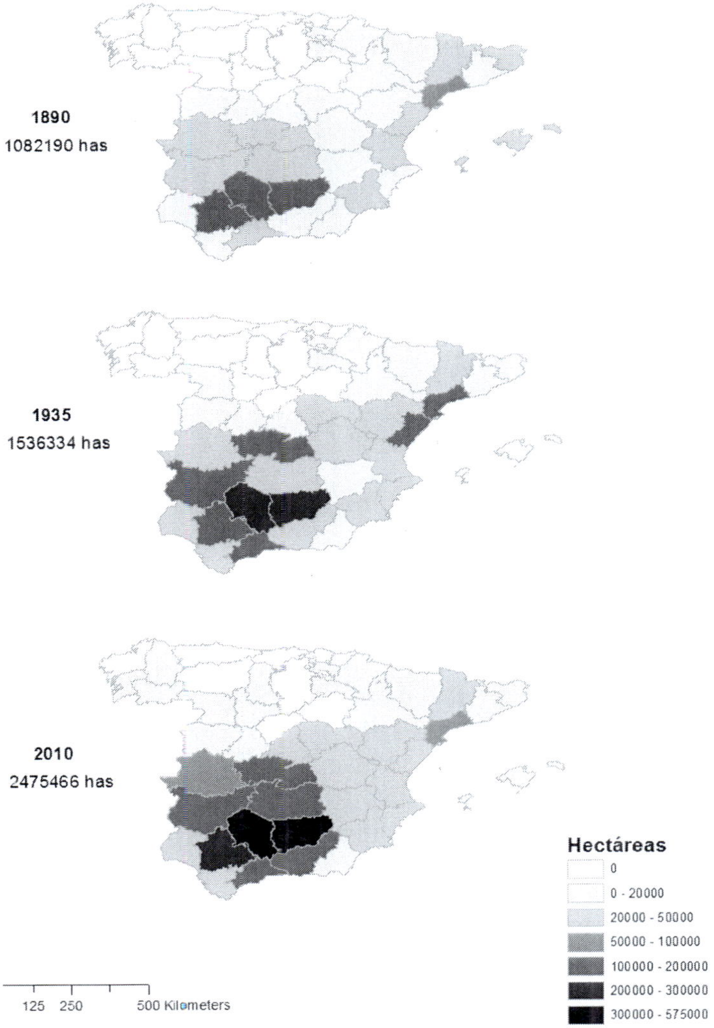


**Figure 1.** Land area occupied by olive groves in Spain, by province: (**a**) in 1890; (**b**) 1935; and (**c**) 2010. Source: [13].

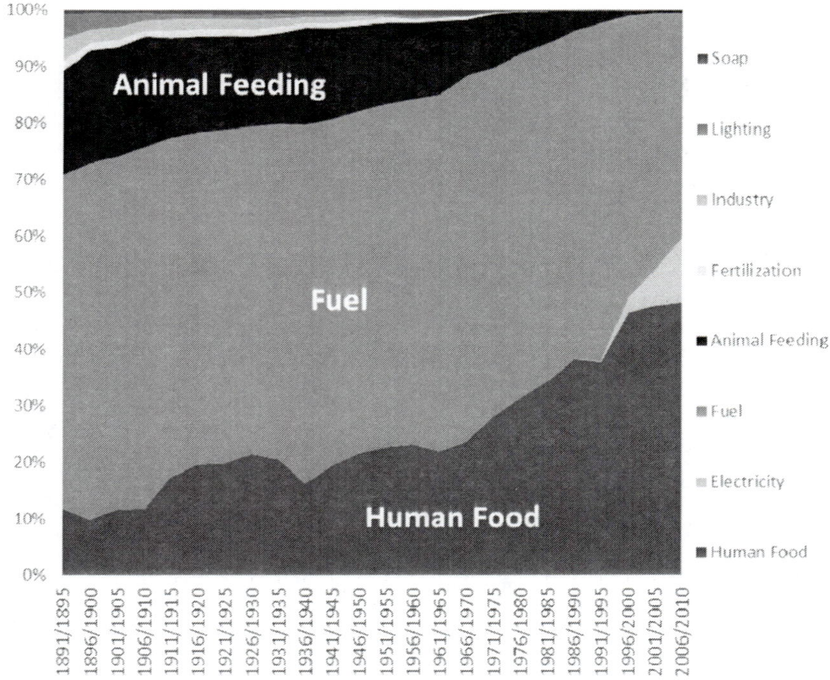


**Figure 2.** End uses of olive products in Andalusia. Source: authors' own elaboration based on [11,13].

A further, and probably more dramatic, change came about as a result of the expansion of the irrigation in olive-growing, mostly in terms of deficit irrigation. Deficit irrigation means applying a limited amount of water which is below the trees' requirement for maximum transpiration and yield. At the end of the 19th Century, barely 70,000 hectares of olive groves were irrigated. In the first half of the 20th century, this figure remained stable at around 100,000 hectares. By the late 20th century this figure had rocketed, reaching the figure of 452,877 hectares by 2009 [19].

The increased use of industrial machinery, synthetic fertilization, and irrigation explains the rise in olive production, mainly through the increased use of external inputs from the mid 20th century onwards. In Andalusia, the average production of olives per hectare increased from 747 kg $ha^{-1}year^{-1}$ in 1896/1900 to 3205 kg $ha^{-1}year^{-1}$ in 2006/2010.

This change also allowed for a substantial rise in the number of trees grown per hectare. Traditional rainfed conditions and systems of fertilization could only sustain low planting density and modest production levels [20]. Between the late 19th century and up until 1960, plantation patterns remained relatively stable throughout Spain, with an average of 80 and 90 trees $ha^{-1}$. Currently, only 37% of the total land area given over to olive groves has 100 or fewer trees $ha^{-1}$, 34% has between 100 and 200 trees $ha^{-1}$ and 28% has over 200 trees $ha^{-1}$ [21].

In this respect, the intercropping of olive trees with other crops has also been reduced: at the end of the 1960s, there were over 300,000 hectares of intercropped olive groves; now, there are just some 191,000 hectares [21]. Furthermore, since the 70s, this same source estimates the number of isolated olive trees scattered among other crops or land uses as rising from almost 4 million in the late 70s to just over 180,000 today.

The history of the olive grove both in Spain as a whole and in Andalusia has been one of steady expansion and major transformations. Olive production is currently seen as a highly productive single crop, and although olives are mostly produced for oil, a significant percentage is also used for the processing of table olives (approximately 5.6% of Spanish olive surface is dedicated to such production). Traditionally (before the transformations of the 19th century) the olive grove played a very different role as the supplier of a wide range of goods and products used for domestic purposes. The process of transformation was brought about by changes in demand for products but also in the management approaches applied to the crop.

In environmental terms and from a historical perspective, this process of intensification required greater input of fertilizers and an increase in all agronomic operations, such as tillage. Case studies have documented that, in the late 19th century, soil losses increased in intensity and the balance of nutrients became negative [14,22]. The industrialization of the sector in the second half of the 19th century led to the mass usage of external inputs, including labor intensive harvesting systems, until the advent of mechanized harvesting in the more modern plantations of the 1990s—which has also generated major problems such as increased soil losses, the contamination of soil and waterbodies, and the loss of biodiversity [1].

# 3. HISTORICAL EROSION RATES IN OLIVE ORCHARDS

Unfortunately, soil erosion in the Mediterranean is not a recent phenomenon and has been widespread since ancient times [23]. As a result of this long-lasting and continued human pressure, present-day degradation of soils has reached severe levels. Clear warning signs, such as completely truncated soils or areas with extremely stony soil, are visible throughout many of the Mediterranean olive-growing areas on slopes. Today's olive farmers have thus inherited a long history of soil erosion acting on their soils. Many of them are confronted with soil properties that have either already changed irrevocably, such as soil depth, or are progressing towards a critical threshold where the soil's response to environmental drivers (for example in terms of infiltration capacity) could change drastically if the remaining fertile soil layer is washed away. In order to design suitable management strategies, it is useful to take into account the history of soil erosion in olive orchards and, ideally, draw lessons from the past in order to prevent future degradation. The study of historical soil erosion dynamics also allows us to evaluate the effect of extreme events, which under Mediterranean conditions can be responsible for more than half of the soil erosion over decadal time [24]. Their low recurrence interval implies that it is difficult to measure them in field experiments, whose record span does not go yet beyond 5–10 years. Determining soil erosion rates over time scales that encompass at least several decades can provide a very much needed insight into long-term trends even though it usually implies greater uncertainty about the absolute accuracy of the measurements, as compared to more controlled experiments.

The evolution of soil erosion in olive orchards is closely tied to their expansion and the changes in soil management. As indicated in the preceding section, soil management practices have significantly changed over the ages. This, together with a rapid expansion over the last two centuries of the extent of olive-growing areas, (Table 1 and Figure 1), has led to important changes in soil erosion rates and its relative contribution to the overall sediment budget on regional scale.

An initial estimate of historical soil erosion rates in olive orchards can be made based on the measurement of the height of the mounds that are formed around the olive trees [25]. This simple technique takes advantage of the permanent character of this tree crop. From a set of measurements of tree mound height in an orchard, the original soil surface at the time of planting can reconstructed and interpolated. The difference with the current surface allows us then to determine the eroded soil volume over the area being measured (Figure 3). If the trees' planting date can be established by interviews or, preferably, with the help of historical documents, soil loss rates can be calculated over a time period of up to almost 300 years, which is the maximum tree age observed in the field so far. Using this technique, we determined [22,25] soil loss rates over time scales of decades to centuries, and reported mean soil loss rates between 61 and 184 t $ha^{-1}$ $year^{-1}$ over 7 different sites near Córdoba (Southern Spain)and between 23 and 64 t $ha^{-1}$ $year^{-1}$ over 3 different sites near Montefrío (Southern Spain).

In order to clarify the relationship between soil erosion and soil management, more studies on historical soil erosion rates are needed. Changes over the last decades have been drastic, and include, for example, the introduction of mechanized agriculture with frequent tillage. Radioisotope tracer techniques are an ideal tool for the study of historical soil erosion and deposition rates on medium-term time scales, and have been successfully applied in different parts of the world. The most widely used isotope is $^{137}Cs$, whose half-life allows us to determine soil erosion rates over a time scale of about 60 years, although other isotopes such as $^{210}Pb$ could be used for longer time scales (*ca.* 100 years). However, up to date, only a few studies have used this technique in fields under olive cultivation. Mabit *et al.* [26] evaluated the potential of this technique in the area of Montefrío (Southern Spain). After extensive validation which used an unexcavated area within a protected archaeological site from Iberian times as a reference area, they concluded that erosion rates in a nearby orchard had reached maximum average values of 19 t $ha^{-1}$ $year^{-1}$ from the 1950's to date, with an estimated sediment delivery downslope of 29%. These figures imply not only an accelerated erosion rate on the hillslope, but also that even on this field's scale, about one third of the eroded soil is redeposited on the footslopes. Similar results have been obtained in other studies on historical soil erosion. For example, in Belgium, a sediment budget in a small river catchment (*ca*. 758 $km^2$) over the last 10,000 years, revealed that 15% of the eroded sediment was still stored as hillslope deposits [27], and a bigger amount as floodplain sediment (68%). In Mediterranean environments under olive cultivation, there is the need to collect more data before a similar sediment budget can be established on an historical time scale.

To our knowledge, only one study exists that integrates soil erosion and sediment transport over larger spatial scales and longer time frames (up to 4000 years) in olive growing areas. Martín-Puertas *et al.* [28] studied sediment deposits in the Laguna de Zoñar, a 370 ha endorheic basin in

Southern Spain. While the land use has varied a lot over this period, the sediment record indicates important anthropogenic activity associated with olive trees since Iberian times (*circa* 500 BC), and at present, the catchment is almost exclusively under olive cultivation. Short periods of increased erosion and sediment delivery to this lake were identified in 650 BC, 300 BC, 300–370 AD, 1350, and during the last centuries two important episodes were identified (1950–1970 and after 1980). While these phases of increased erosion reflect both climatic and human influences, the heavy metal enrichment ratios that indicate an intensification of human activity during the Iberian-Roman Period and the use of fertilizers during the last 50 years are especially significant.

**Figure 3.** Reconstruction of historical soil loss since olive plantation based on surface topography reconstruction with tree mounds.

The field measurements and datings discussed above are crucial for quantifying historical soil erosion rates. However, the modeling of historical soil erosion dynamics is an indispensable tool to better understand the relationship between soil management intensity and soil erosion rates in the past and to extrapolate the limited field data available to a broader area and a higher temporal resolution. One of the major limitations for the application of soil erosion models at historical time scales is originated by the necessary input parameter data and by the limited data set for validation. Based on an extensive study of the documentary sources [22], soil management could be documented in detail over a 250 year period in the Montefrío area (Southern Spain). Eight different periods were distinguished, starting from low-intensity management, with two light animal-drawn tillage operations per year, to high-intensity management, with at least two passes per year with the moldboard plow and between two and four (shallower) passes with the harrow. Based on this information, key parameters such as vegetation cover, plowing depth, or the duration of periods with bare soil could be determined. Rainfall erosivity was reconstructed from a combination of rain gauge measurements and historical documents. Finally, this input dataset allowed us to run a combined analysis using a water erosion model (based on the Revised Universal Soil Loss Equation or RUSLE) calibrated for olive orchards according to the methodology established by [29], combined with a tillage erosion model, which was specially designed for this study and applied on three different sites. The model reproduced the temporal patterns in soil erosion rates and showed considerable historical variation. Figure 4 represents the average soil erosion rates of the three sites: between 16 and 84 t $ha^{-1}$ $year^{-1}$ for water and between 3 and 28 t $ha^{-1}$ $year^{-1}$ for tillage. The results showed that even before the advent of mechanized agriculture, a period often associated with sustainable land use, soil loss rates were already relatively high: 32 t $ha^{-1}$ $year^{-1}$. The model also pinpointed a particular period between 1935–1970 with considerably lower soil erosion rates, as cereals were grown among the olive trees due to the autarchic economic regime after the Spanish Civil War. Finally, after the introduction of mechanized agriculture in the 1970s, combined soil erosion (water plus tillage) rates almost tripled. The model results also allowed us to distinguish between the relative significance of water and tillage erosion processes, the former being identified as the most important. It is also important to note that, since the late 2000's, a combination of agro-environmental regulations coupled with the higher cost of fuel has resulted in a reduction of tillage operations, as well as of the plowing depth, which resulted in a markedly decreased rate of tillage erosion.

There are no indications in traditional sources of extensive erosion damage in olive orchards prior to the expansion of olive cultivation in the mid-18th century [11]. This can be explained by the low intensity of the tending systems applied and extensive vegetation ground cover permitted by these systems. However, the situation has changed during the last 250 years, in which this crop has rapidly expanded all over the Southern part of Spain. Although soil management intensity was relatively low before the start of mechanized agriculture in the last decades, the intensity of soil erosion has been shown to be unsustainable since as early as the late 18th century. It is clear that there has been an important time-related variability of soil erosion rates over time due to direct feedback between soil management and the intensity of soil erosion. Historical data has clearly shown that during periods of increased soil cover, soil erosion rates are considerably reduced, for example during the 1935–1970 period in the area of Montefrío, which was intercropped with cereals. Especially in the last decades, however, soil erosion under olive cultivation has been especially intense and rates have been shown to have tripled, due to a cultivation system based on the complete elimination of the vegetation ground cover among the olive trees all year round. More quantitative field data and dating will be needed before the results of these case-studies can be extrapolated to broader areas with reasonable certainty. In addition, the contribution to soil loss of processes such as gully erosion, shown to be important in many Mediterranean areas as well as in olive growing areas in Andalusia [30], is yet unknown on a historical time scale.

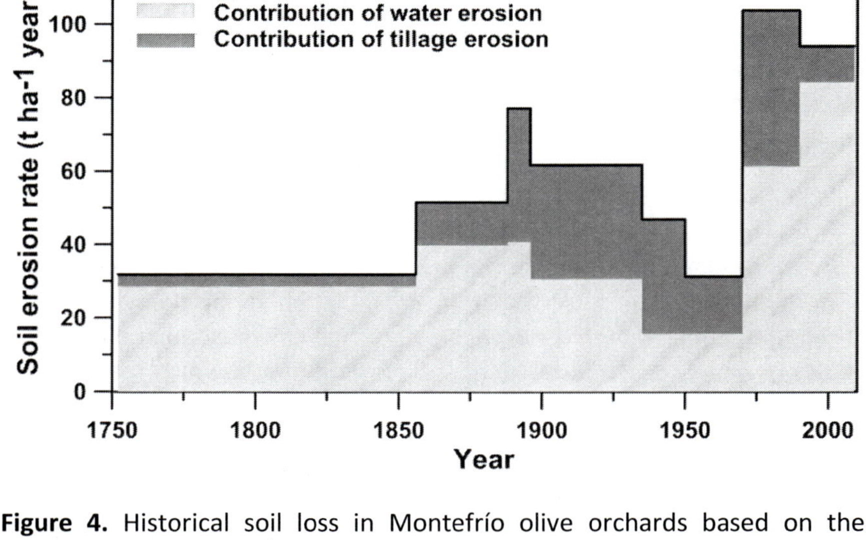


**Figure 4.** Historical soil loss in Montefrío olive orchards based on the average of three study sites [22].

## 4. APPRAISAL OF CURRENT EROSION RATES AND SOIL QUALITY IN OLIVE ORCHARDS IN ANDALUSIA

There are basically two major approaches to soil management in olive orchards in Andalusia, and broadly speaking, all over the world: those which eliminate the ground cover in the area outside the projection of the olive canopy (which is achieved through plowing, the use of herbicides or a combination of both) and those which provide ground cover, usually achieved using a cover crop (seeded or spontaneous), sometimes combined with a mulch of pruning residues. In the latter case, control of the growth of the cover crop in spring is required to prevent competition for soil water with the trees, which are still mainly rainfed or, when irrigated, are under deficit irrigation. Although there are variations within each of these two main approaches (for instance, the number of times plowing is applied, whether a herbicide is or not used for bare soil management, the extension and type of the cover crop and the date of its control and the method used: either mechanical, animal or chemical), the major differences in soil management and its impact on soil properties are related to the presence or not of ground cover beyond the one provided by the olive tree canopy [31]. In Andalusia, we have found two kinds of studies measuring erosion rates in olive groves. Some are based on runoff plots and are aimed at comparing the effects of different soil managements on soil losses under the same environmental conditions (rainfall, topography, soil type) at hillslope scale. The other kind of studies are based on measurements taken on a small catchment scale (7 to 12 ha). The latter provide an integrated measurement of the balance between the different erosion processes and the deposition in the lower areas of the catchments, but are not conceived to evaluate the impact of different soil management under controlled conditions. Table 2 summarizes the results of these experiments. Hillslope-scale, plot-based experiments usually show unsustainable erosion rates when bare soil management (continuous tillage, CT, or no tillage with bare soil, NT) is used, which is in line with the observations of long-term erosion rates based on indicators such as olive mounds [22] or radionuclide content on eroding profiles [26]. Although NT and CT are different soil management methods, they result in comparable water erosion rates. It is also apparent from Table 2 that erosion rates under the same conditions and with the use of a cover crop are greatly reduced, usually below or closer to tolerable erosion rates. Within the context of erosion studies in olive orchards, tolerable erosion rates usually range from 10 to 12 t ha$^{-1}$ year$^{-1}$ [32] following the approach recommended for tree crops by the literature. The results shown in Table 2 on small catchment scales also indicate relatively high erosion rates, with the exception of one experiment, where a combination of cover crops and rocky outcrops (and a notably dryer rainfall regime) within the catchment resulted in low erosion rates. These results are consistent with the concern about high erosion rates and degraded water quality due to offsite contamination (which is indirectly related to high sediment load from the cropping areas) in olive growing areas in Southern Spain [1].

**Table 2.** Measured soil losses by water erosion on hillslope (plots) and catchment scale experiments in olive orchards. NT means bare soil with herbicide use and no tillage, CT bare soil by tillage, CC cover crop soil management.ources: [1] [33], [2] [34], [3][35,36,37], [4] [38], [5] [39], [6] [40], [7] [41], [8] [42].

| Experiment | Soil Management | Average Soil Loss t $ha^{-1}$ $Year^{-1}$ | Number of Years | Slope Length (m) | Steepness of Slope (%) | Plot Area ($m^2$) |
|---|---|---|---|---|---|---|
| Conchuela I [1] | NT | 6.9 | 5 | 12 | 13.4 | 72 |
| | CT | 2.9 | 5 | 12 | 13.4 | 72 |
| | CC | 0.8 | 5 | 12 | 13.4 | 72 |
| Benacazon [2] | CT | 29.6 | 6 | 60 | 11 | 480 |
| | CC | 58.9 | 3 | 24 | 13.4 | 288 |
| Conchuela II [3] | CT | 5.6 | 2 | 24 | 13.4 | 288 |
| | CC | 5.6 | 2 | 24 | 13.4 | 288 |
| Lanjarón I [4] | NT | 25.6 | 2 | 24 | 30 | 480 |
| | CT | 5.7 | 2 | 24 | 30 | 480 |
| | CC | 2.1 | 2 | 24 | 30 | 480 |
| Lanjarón II [5] | NT | 25 | 2 | 22.5 | 30.5 | 480 |
| | CT | 8.5 | 2 | 22.5 | 30.5 | 480 |
| | CC | 6.5 | 2 | 22.5 | 30.5 | 480 |
| **Experiment** | **Soil Management** | **Average Soil Loss t $ha^{-1}$ $Year^{-1}$** | **Number of Years** | **Maximum Flow Length (m)** | **Average Slope (%)** | **Catchment Area ($m^2$)** |
| Conchuela [6] | CT and CC | 16.1 | 5 | 270 | 9 | 80,000 |
| Setenil [7] | NT and CT | 9.3 | 6 | 340 | 9.5 | 67,000 |
| Puente Genil [8] | CC | 1.8 | 5 | 150 | 15 | 61,000 |

One major question is how to extrapolate these experimental records to the 1.5 Mha of olive groves in the region. This has been done through use of RUSLE [43] to simulate the different conditions (of slope, soil and climate conditions and soil management) present in olive orchards in the region. Gómez *et al.* [29] proposed a methodology for calibrating the RUSLE factors for different conditions of olive orchards in the region. The prediction capabilities of this approach have been evaluated with long-term erosion rates [22] with a good level of accuracy. Marín [44] using most of the experimental data summarized in Table 2 found good correlation with a RMSE of 4.6 t $ha^{-1}$ $year^{-1}$ and a model efficiency (Nash and Sutcliffe) equal to 0.82. Using this methodology, Gómez and Giráldez [32] calibrated RUSLE on the scale of a "comarca" (roughly, the Spanish equivalent of a county) using the rainfall, slope and soil properties available for the K, R, L and S factors of RUSLE.Figure 5 and Table 3 summarize the main results of this study, including the definition of those factors. Figure 5 summarizes the erosion losses caused by water on a hillslope scale in the different areas in Andalusia according to different scenarios. Scenario 1 is the scenario which assumes bare soil management, which was the standard situation stemming from the widespread introduction of mechanization and the use of herbicides in the late 1950's, up until the introduction of agro-environmental regulations (national or European) to prevent soil erosion in the mid 2000's. Scenario 2 is the present situation according to expert analyses [32], which points towards a moderate adoption of cover crop use. Scenarios 3 and 4 are hypothetical scenarios considering more appropriate soil management techniques (basically, an expansion of cover crop use), especially in the areas with higher erosion risk, even if this leads to a reduction in yield due to competition for soil water between the olive trees and the cover crops.

Model predictions depicted in Figure 5, and summarized in Figure 6, indicate that unsustainable erosion rates of over 10 t $ha^{-1}$ $year^{-1}$ are still common in the region. It is worth noting that this may have been a common situation in the region since the late 1950's due to a combination of cultivation in sloping areas, a relatively high erosivity and unsuitable soil management. Since the situation in the decades prior to the late 1950's must have resembled an intermediate situation between Scenarios 1 and 2, this model analysis indicates that the unsustainably high erosion rates must have been common in most traditional olive growing areas in the region for a long time. This is consistent with the findings of Infante-Amate and González [14], which indicate a severe P deficiency in traditional olive groves during the 19th century, which can only be explained by P losses from soil erosion. The same observation of symptoms of unsustainable erosion rates was made by experts during the early 1960s in olive orchards areas in Southern Spain which were still not mechanized [52]. Finely-tuned soil management techniques specifically oriented to soil conservation, with increased use of cover crops (including their control later in spring), which can even result in a decrease in olive yield [32], might be necessary to achieve "tolerable" rates (below 10 t $ha^{-1}$ $year^{-1}$) of erosion. In addition, even in the most sloping (some above 35%) and degraded areas, measures based on the abandonment of olive cultivation to allow natural reversion to

Mediterranean forest or other agro-forestry systems [48] should be promoted (Figure 6).

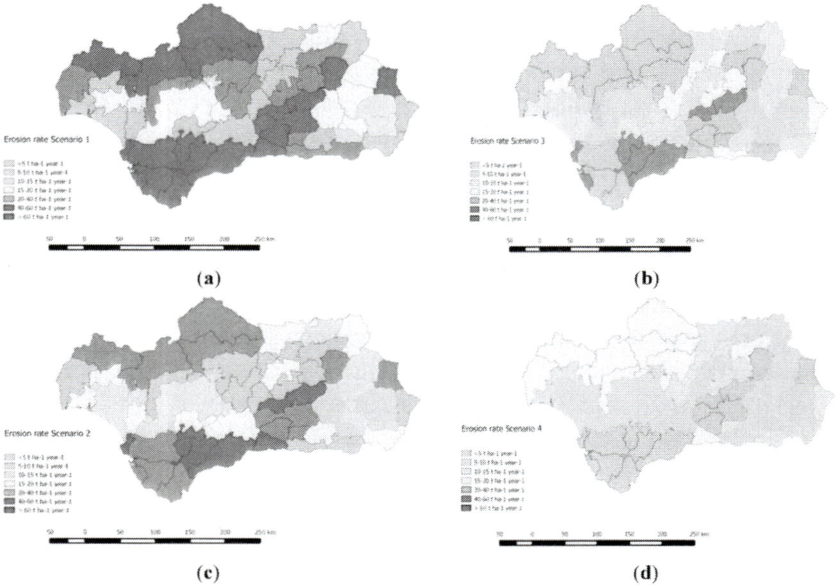


**Figure 5.** Maps of average erosion rates on the hillslope scale in olive orchards by "comarca" (county) according to Scenarios 1 (**a**), 2 (**b**), 3 (**c**) and 4 (**d**) respectively. Scenario 1 assumes bare soil management ($C_1$); Scenario 2 represents the current situation with a moderate adoption of cover crop use ($C_2$); Scenarios 3 and 4 ($C_3$ and $C_4$) are hypothetical scenarios considering more appropriate soil management techniques, especially in those areas with higher erosion risk. See also Table 3.

Figure 7 presents a summary of key soil properties of the top 10 cm of soil found in different studies of olive orchards in Andalusia. Although there is a wide variability within each class resulting from the heterogeneity of the soil conditions dealt with by the different studies, a trend towards lower phosphorus and organic matter content and aggregate stability is apparent in olive orchards managed with bare soil, as compared to those with cover crops. In controlled experiments, the correlation between soil management and soil quality indicators is much clearer. For instance, Gómez *et al.* [33]

found in a 5 year experiment, a clear correlation between the different erosion rates induced by different soil management techniques (cover crop, bare soil by tillage and bare soil by herbicide) and top soil properties: higher erosion rates and worse soil quality indicators are found as a consequence of bare soil management as compared to those managed with a cover crop, the latter producing the best indicators, and tillage presenting intermediate results. The results in Figure 7 also indicate a significant decrease in organic matter and aggregate stability of top soil in olive orchards as compared to natural areas.

## 5. SOIL WATER BALANCE AND OLIVE YIELD

Olive yield, as it is the case of other crops grown in arid and semiarid conditions, responds non-linearly to seasonal evapotranspiration (ET). Moriana *et al.* [54] measured the total ET from soil water content readings using neutron probe measurements. In an adult olive orchard located in Cordoba (Southern Spain) submitted to different irrigation regimes (from fully irrigated to rainfed) they demonstrated that there is a polynomial correlation between seasonal (January to December) olive orchard ET and olive yield, with a diminishing increase in olive yield for higher ET values.

With the objective of determining a robust relationship between current olive tree transpiration and olive yield, we decomposed the two components of ET for the experiments reported by [54] using a process-based water balance model designed for simulating the components of the hydrologic balance in olive orchards, WABOL [55]. We did this by simulating the meteorological conditions for the years considered (1996–1999) from daily records, and the three irrigation treatments (fully irrigated, 50% of required irrigation, and rainfed) indicated by [54], hereafter referred to as Cordoba. To achieve this, WABOL was modified to include specific modules for drip irrigation and evaporation from the soil surface in the area wetted by the dripper. To expand the dataset provided by [54], we also modeled the components of the hydrological balance for orchards under rainfed conditions in two traditional olive production areas in Andalusia considering those years for which yield records were available. One the sites was an area showing very low yields, close to the town of Obejo, and another one offering medium yields, close to Baena. For these two areas, comprehensive descriptions of the orchards and average yields for three consecutive years were available in [49] for Obejo and [50] for Baena. In the Obejo and Baena cases, for which only a three-year record was available, we used the trianual average annual yield, while in the Cordoba experiments we used the biannual average annual yield, orchard evapotranspiration and orchard transpiration data. Due to alternate bearing, the use of biannual averages is standard in agronomic studies of olive yield. In this approximation, we assumed that the model properly determines olive transpiration and soil evaporation, an assumption already made in previous studies [56] using the same equations implemented in WABOL. Table 4 provides a summary of the simulations performed as well as some key orchard variables.

**Table 3.** Summary of the determination of RUSLE factors for the erosion maps by county and management scenario. Sources [29,43,45,46, 47,48, 49,50,51]. $C_1$ to $C_4$ refer to management scenarios 1 to 4.

| Variable | Sources | Methodology |
|---|---|---|
| Rainfall erosivity, R | [45] | Interpolated using R values in the stations indicated in [45] averaged by "comarca" |
| Soil erodibility, K | [43,46,47] | Calculated following [43] from the soil properties of the main soil types in each "comarca" (taken from [4]), weighted average by the extension of each soil type within each "comarca" taken from [46]. |
| Slope steepness S | [48] | Calculated as from [43] for each soil class within each "comarca" weighted average according to the extension of each soil class indicated in [48]. |
| Slope length, L | [49–51] | Calculated from $L = 712.78\ S^{-0.6829}$, with S slope steepness. Imposing a maximum L value of 240 m based on field observations indicated in [49–51]. |
| Cover and management factor, C | [29] | According to [29], including stone ground cover in each "comarca" and average ground cover of olive trees of 35%.<br>$C_1$: Bare soil by plowing and/or herbicide.<br>$C_2$: Orchards above 15% slope using a cover crop covering 25% of the ground, mowed in mid-spring. Below 15% plow following contour lines.<br>$C_3$: Orchards below 10% slope plowed following contour lines. From 10% to 30% with a cover crop as in $C_2$ but covering 33% of the ground. Above 30% slope as in $C_2$ but covering 50% of the ground.<br>$C_4$. Orchards below 5% plowed following contour lines. From 5% to 20% slope as in C2 but covering 33% of the ground and following contour lines. From 20% to 30% covering 50% of ground cover and disposed following contour lines. Above 30% slope as in $C_2$ but providing full ground cover |

Figure 8a depicts the predicted ET values against the reported yields, indicating in each case the orchard conditions. On the same graph, we have represented the relationship between ET and orchard olive yield proposed by [54] with a red line ("Moriana *et al.*, regression"). It is noticeable how the two graphs overlap in the region corresponding to the simulated points for the Cordoba treatments, indicating how the model accurately predicts the ET measured by [54]. It is also noticeable how the points corresponding to the Obejo and Baena simulated treatments depart slightly from the section of the regression determined by [54], a section of lower yields for which [54] there were no experimental records. Nevertheless, the overall dataset follows a similar non-lineal trend of decreasing yield with decreasing ET.

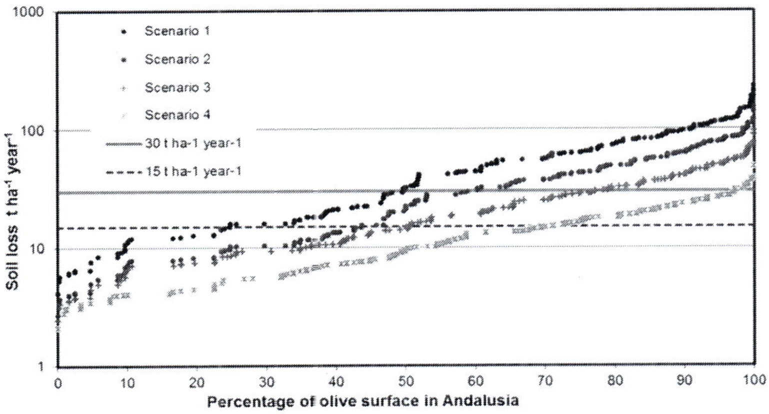


(a)

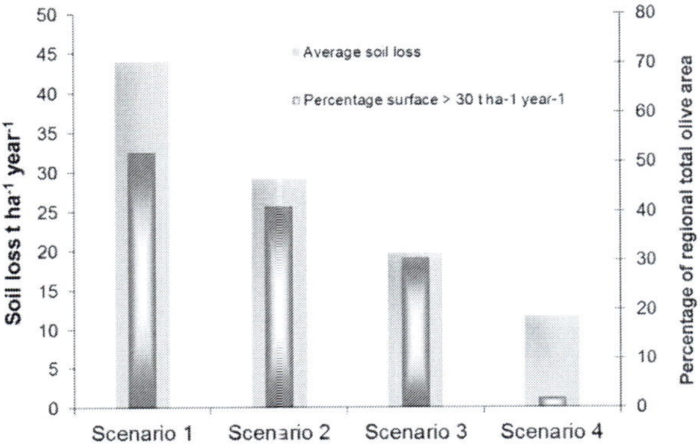


(b)

**Figure 6.** (**a**) Cumulative distribution of olive orchard surface in relation to erosion risk according to the four scenarios described in Table 3; (**b**) Average erosion rate and percentage of olive orchards above 30 t $ha^{-1}$ $year^{-1}$. Scenario 1 assumes bare soil management; Scenario 2 represents the current situation with a moderate adoption of cover crop use; Scenarios 3 and 4 are hypothetical scenarios considering more appropriate soil management techniques, especially in the areas with higher erosion risk. See also Table 3.

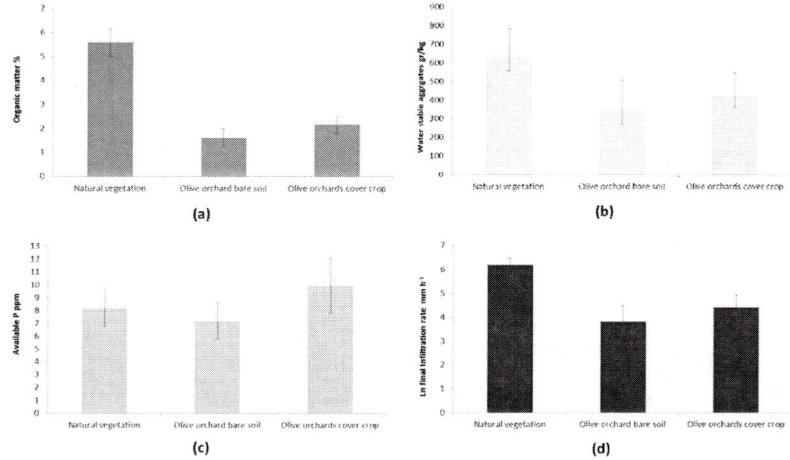


**Figure 7.** Summary of key soil quality indicators in olive orchards, under bare soil or cover crop management, and nearby natural vegetation areas in the top 0–10 cm of the soil. Sources: [33,38,49,50,51,53].

**Table 4.** Summary of WABOL simulations. ** Irrigated 5 days a week for six hours from 15 April to 10 October using 4 emitters of 8 liter hour$^{-1}$ per tree.

| Place | Management | Tree Density (Tree ha$^{-1}$) | Soil Texture, Depth (cm) | Seasonal Years Simulated |
|---|---|---|---|---|
| **Cordoba** | Fully irrigated ** | 278 | Sandy-loam, 220 | 1997–1999 |
| **Id.** | Rainfed, tilled | 278 | Id. | Id. |
| **Obejo** | Rainfed, tilled | 150 | Sandy-loam, 80 | 2003–2005 |
| **Id.** | Rainfed, cover crop | 115 | Id. | Id. |
| **Baena** | Rainfed, tilled | 100 | Clay-loam, 120 | 2003–2005 |
| **Id.** | Rainfed, cover crop | 100 | Id. | Id. |

It is apparent in Figure 8a how there is a very good fit between olive yield and orchard seasonal T, $R^2$ = 0.98 and RMSE = 0.60 t year$^{-1}$, with a diminishing increase for higher T values as observed by [54] for the aggregated T and E values. This regression can be used to convert the olive transpiration modeled into expected olive yields (assuming that no other environmental or agronomical limitations appear). It will also allow us to evaluate the impact of changing climate or management conditions on yield changes, caused by variations in the water available for olive tree transpiration.

# 6. IMPACT OF SOIL LOSSES ON OLIVE PRODUCTIVITY

The impact of soil loss on crop productivity mainly occurs in terms of the limited availability of nutrients and water to the crops, but also through physical damage such as root exposure or burial in depositional areas [57]. Several methods can be used to assess the effect of soil loss on crop productivity, each of which presents specific advantages and limitations. Because of the complexity of the task, the use of simulation models has become very common [57] in order to reproduce the interactions of different factors, which could compensate or exacerbate the effect of soil losses. In fact, the study on historical erosion rates in Montefrío presented in Section 3 indicated how yield had increased over the last 200 years due to the improvement of agronomical techniques, while suffering the loss of approximately 40 cm of the original 120 cm estimated original top-soil.

It has been recognized that water availability is the major production constraint for olive cultivation [54]. For a crop which is mostly rainfed, especially in steep areas under arid and semi-arid conditions, the role of soil as a water reservoir for the plant during the dry season is critical, and should play a major role in determining the potential productivity of an olive orchard (assuming that other factors such as nutrient availability or pests and diseases are not limiting productivity). We evaluated the impact of erosion on productivity of rainfed olive orchards using a water balance model, WABOL [55], and the relationship between olive annual transpiration and annual olive yield presented in Figure 8b. We considered two examples which were representative of the two different situations of olive orchards summarized in Table 5.

Case study 1, Obejo, is representative of rainfed orchards in very steep areas with shallow soil profiles, coarser textures and high stone content. Case study 2, Cordoba, is representative of olive orchards cultivated in the rolling landscape of the Guadalquivir river valley on deeper soils with finer texture, fewer stones and better water-holding capacity. For each soil depth and case study, 30 years were simulated, the ones corresponding to the 1971–2000 period. The outputs for each simulation were used to generate 3D graphs with soil depth and annual rainfall on the X and Y-axes respectively, and yield (calculated using the expression shown in Figure 8b) on the Z-axis. These surfaces are represented in Figure 9. It is apparent from those surfaces that, given the different hydrologic conditions of the orchards, the response of yield to different soil depth and annual rainfall differs, demonstrating different sensitivities to a decrease in soil depth by erosion.

For easier interpretation, Figure 10 a,b depicts curves corresponding to specific rainfall values of the surfaces shown inFigure 9. It is apparent that in the orchards represented by the Cordoba scenario case study there is lower yield sensitivity to the decrease of available soil depth, unless these values reach an inflexion point close to 60 cm. We attribute this to the relatively high water holding capacity of the soils simulated in this scenario, which is common in orchards located in the less steep areas of the region. As a result, soil depths of over 60 cm seem to be enough to maintain a water storage capacity for the dry season which is capable of supporting stable yields during most years. The simulations for this typology of olive orchard, based on larger crown canopies for optimum use of the higher water storage capacity of the soil, are extremely sensitive to a reduction of average rainfall values, especially below 400 mm. A contrasting situation occurs for the Obejo case study, which is representative of many traditional olive growing areas in mountainous parts of Andalusia. In the Obejo case, there is a decrease in potential yield productivity due to the reduction of soil depth even when the soil profile is still relatively deep. This rapid response of potential yield to soil depth reduction is a consequence of the low water holding capacity of the soil, which implies that the water front infiltrates deep below the olive root zone into the relatively permeable parent material. Low water holding capacity is a result of a coarser texture and a higher stone content, and can be expected in soils in olive orchards in the mountainous areas of Andalusia. Under the conditions of the Obejo case study, a much deeper soil is required to store most of the infiltrated rainfall for use by the olive tree in the dry season. Simulations predictions show how this kind of olive orchards, which have a relatively small crown canopy and a reduced tree density, have some yield response to low rainfall values, around 200 mm. This low rainfall depth resulted in the collapse of the orchards as simulated by the Cordoba scenario. If we connect the current soil depth in the Obejo case study with the cumulative soil losses in a similar environment such as Montefrío [22], it is apparent that the orchards have already accumulated a significant loss of potential yield due to losses of soil profile since the time olive cultivation started in the area between the 18th and 19th centuries. It is also interesting to note how the decrease in potential yield in the Obejo case study is gradual, without an apparent collapsing point in which a dramatic decrease in yield might be perceived. On the other hand, for the Cordoba site, the drop in yield once the soil depth falls below 50 cm is much more pronounced, which indicates that, at that depth of soil, there is a significant decrease in water available to the tree.

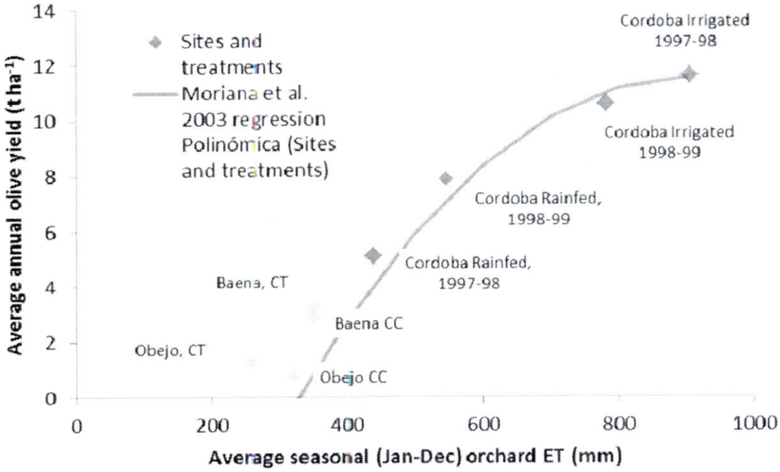


(a)

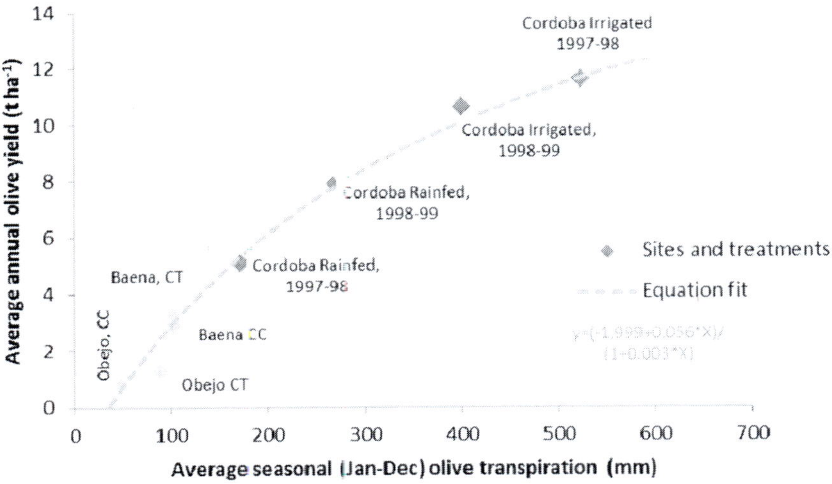


(b)

**Figure 8.** (**a**) Correlation between predicted ET and measured yield for the different simulated conditions (blue [54], brown [50] and green [49] dots indicate sources of the yield data). The red line indicates regression based on field measurements between seasonal ET and olive yield by Moriana *et al.* [54]; (**b**) Correlation between predicted T and measured yield for the different simulated conditions (blue dots have the same meaning as the colored dots in Figure 8a. Also, the fitted regression between seasonal T and olive yield.

**Table 5.** Summary of simulated conditions with WABOL. Figures of soil depth in black indicates current soil depth. For soil depths above the current one, the $\theta_{Field\ Capacity}$was increased by two (0.02) units, and CN numbers decreased by two units to consider better soil quality. For those below the current depth, the $\theta_{Field\ Capacity}$ was decreased by two (0.02) units and CN numbers increased by two units to consider worsened soil quality conditions. The orchard characteristics were taken from [49] and [50].

| | Soil Depths Simulated (cm) | Textural Class | Stoniness % | $\theta_{Field\ Capacity}$ ($m^3\ m^{-3}$) | $\theta_{Permanent\ Wilting\ Point}$ ($m^3\ m^{-3}$) | Slope % | Tree Density (Olive $ha^{-1}$) | Tree Crown Volume ($m^3$) | Plant Leaf Area Density ($m^2\ m^{-3}$) | $CN_I$-$CN_{II}$-$CN_{III}$ |
|---|---|---|---|---|---|---|---|---|---|---|
| Obejo | 20, 40, 80, 120, 180 | Sandy-loam | 33 | 0.21 | 0.09 | 30 | 139 | 24 | 1.15 | 65-81-91 |
| Córdoba | 20, 40, 80, 120, 180 | Clay-loam | 3 | 0.40 | 0.20 | 11 | 139 | 31 | 1.45 | 68-84-94 |

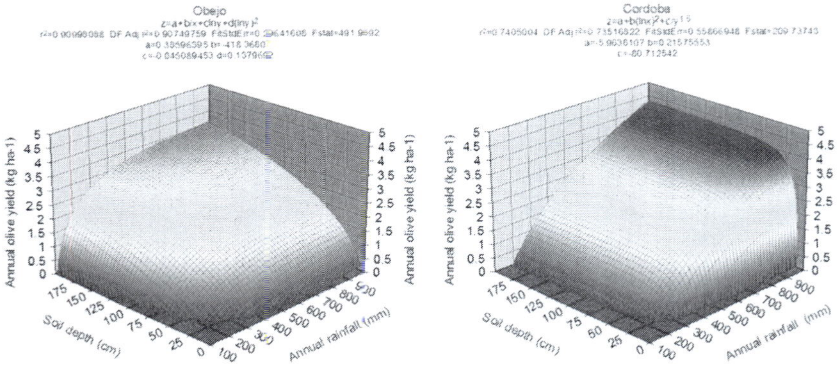


(a) (b)

**Figure 9.** Relationship between potential yield (*Z*) and annual rainfall and soil depth (*X*, *Y*) for the Obejo (**a**) and Cordoba; (**b**) case study.

Based on the relationships summarized in Figure 10a,b, it is possible to convert the cumulative olive yield losses for both examples using current prices (€400 per ton of olives for oil processing), into Euros per hectare. Figure 11 depicts both relationships. It is apparent how for these examples—which can be representative of two contrasting and common situations in the region—productivity losses by erosion due to the decrease in water availability have already significantly affected the productivity of olive orchards situated in mountainous areas in the regions. The orchards represented by the Obejo case study, which have the lowest productivity rate as well as the highest unit production costs in the region have already suffered from a decrease in productivity due solely to the effect of erosion on soil water storage capacity, as compared to the orchards located in the valley (e.g., the Cordoba case study). Since access to irrigation is not feasible in most sloping areas, continued degradation will only further decrease the potential productivity of the olive trees or any other future land use in these areas (e.g., grazing or forestry uses). It has also diminished its relative competitiveness compared to the olive producing areas in the best conditions, e.g., Cordoba case study. Figure 11 also provides an approximation to the order of magnitude of economic losses that can be attributed to the decreasing soil depth for both case studies. For the Cordoba example, these losses are, in practical terms, negligible since the current soil depth is still in the non-response plateau. Although other on-site erosion damage, such as reduced accessibility to the fields, are not included in this evaluation, it is an interesting hint that might help to understand the low impact of disseminating soil conservation measures among these farmers. Many of these efforts have highlighted both in documents and technical sessions their impact on the prevention of yield losses, which, as our analysis indicates, can take a long time to have a significant impact in economic terms, even if no agronomic improvements are assumed. The Obejo case study is a contrasting scenario, and there is a significant impact of cumulated loss of soil depth, at an estimated 200 € $ha^{-1}$ $year^{-1}$. The simulated accumulated soil depth required to achieve that impact, some 40 cm, has been actually documented in a similar area by [22]. In this area, Montefrío, that loss of 40 cm of soil occurred over a long time span of approximately 200 years as erosion rates, although relevant, were lower than the current ones. This has probably contributed towards dampening the link between reduction of soil depth and productivity. Even at current erosion rates of between 0.5–3 mm $year^{-1}$, yearly variations in yield due to decreased productivity can not be perceived. Only by comparing periods with enough time lag, and including model analysis, can we provide indication of this decrease in productivity, isolating the effect of other variables that could mask that relationship (climate, pests and diseases, improvements in management, *etc.*)

Current predictions of climate conditions for the next decades in the olive growing areas in Andalusia predict a significant decrease in annual rainfall, as well as an increase in temperature and evapotranspiration (Table 6). These scenarios will result in an increased pressure on water resources and the need to make optimum use of rainfall water for olive cultivation. For those situated in the mountainous areas in the region, maintaining the current soil depth is critical since it is apparently the only viable alternative to maintaining an already limited productivity. This might require a change in soil management: to provide extensive ground cover during most of the year (Scenarios 3 or 4 in Figure 5 and Figure 6). This in turn might have a notable impact on yield but simultaneously have the effect of mitigating the silent decrease in potential yield due to loss of soil water storage capacity, as indicated in Figure 10. For the orchards located in better soils and less sloping areas, the major impact on potential yield seems to be due to the decrease of rainfall rather than decrease of soil water storage (Figure 11). This does not mean that there is no need to curb the already high erosion rates in these areas, but that the policy approach should be different in order to have a maximum impact among key stakeholders, especially farmers. One major feature is that the main reason for encouraging soil conservation measures in these areas, represented in a simplified way by the Cordoba case study, should be the limitation of off-site erosion damage (such as offsite contamination or damage to infrastructures). The greater sensitivity of yield to a decrease in rainfall supply as compared to the case of the more mountainous areas, Figure 9, suggests that these erosion control techniques should be carefully aimed at minimizing the risk for water competition with the tree, due to the to the increase of transpiration by cover crops. For this reason, the selected use of vegetation, such as buffer strips, combined with inert mulches, might provide a compromise between soil conservation and productivity.

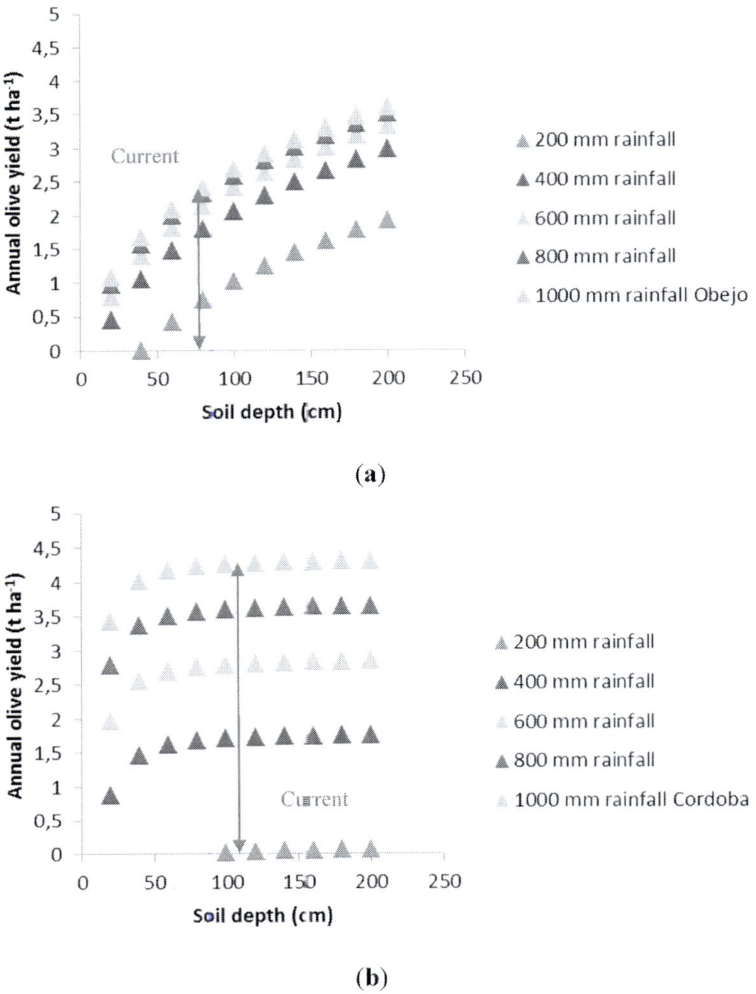


**Figure 10.** Relationship for given annual rainfall between potential yield (*Z*) and soil depth (*X*) for the Obejo (**a**); and Cordoba (**b**) case study.

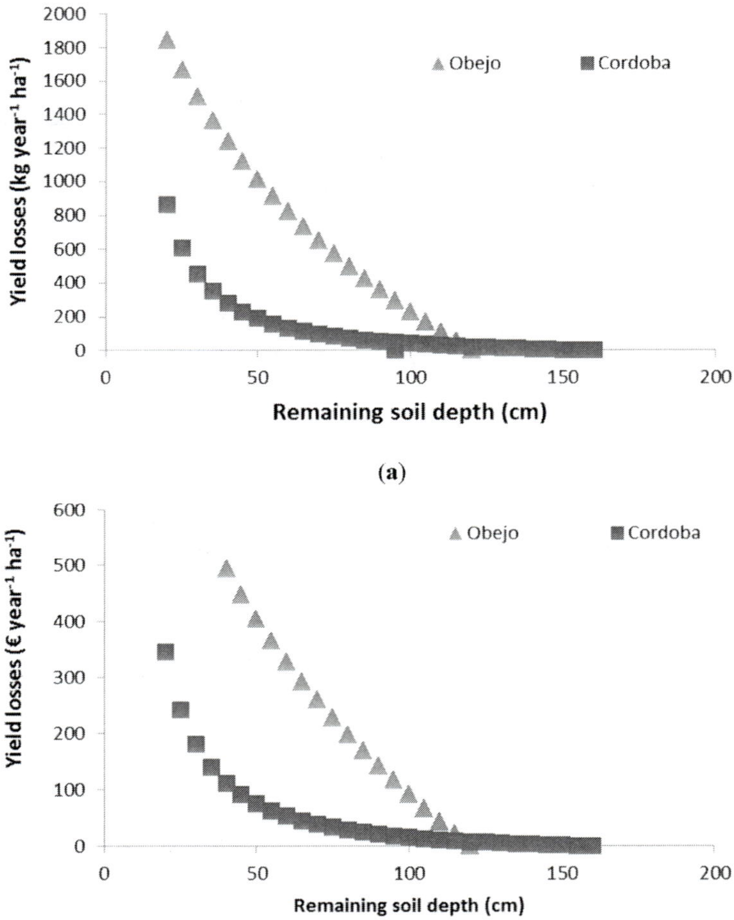


(a)

(b)

**Figure 11.** Relationship between cumulative yield (**a**) economic losses (**b**) and remaining soil depth for the Obejo and Cordoba scenarios. Zero economic losses correspond to the estimated original soil depth prior cultivation. The arrows in Figure (**b**) illustrate the different impact on yield and economic return for a similar remaining soil depth.

Current predictions of climate conditions for the next decades in the olive growing areas in Andalusia predict a significant decrease in annual rainfall, as well as an increase in temperature and evapotranspiration (Table 6). These scenarios will result in an increased pressure on water resources and the need to make optimum use of rainfall water for olive cultivation. For those situated in the mountainous areas in the region, maintaining the current soil depth is critical since it is apparently the only viable alternative to maintaining an already limited productivity. This might require a change in soil management: to provide extensive ground cover during most of the year (Scenarios 3 or 4 in Figure 5 and Figure 6). This in turn might have a notable impact on yield but simultaneously have the effect of mitigating the silent decrease in potential yield due to loss of soil water storage capacity, as indicated in Figure 10. For the orchards located in better soils and less sloping areas, the major impact on potential yield seems to be due to the decrease of rainfall rather than decrease of soil water storage (Figure 11). This does not mean that there is no need to curb the already high erosion rates in these areas, but that the policy approach should be different in order to have a maximum impact among key stakeholders, especially farmers. One major feature is that the main reason for encouraging soil conservation measures in these areas, represented in a simplified way by the Cordoba case study, should be the limitation of off-site erosion damage (such as offsite contamination or damage to infrastructures). The greater sensitivity of yield to a decrease in rainfall supply as compared to the case of the more mountainous areas, Figure 9, suggests that these erosion control techniques should be carefully aimed at minimizing the risk for water competition with the tree, due to the to the increase of transpiration by cover crops. For this reason, the selected use of vegetation, such as buffer strips, combined with inert mulches, might provide a compromise between soil conservation and productivity.

**Table 6.** Predicted variations in average rainfall, ETo and temperatures from a control period (from 1971 to 2001) compared to a future period (from 2068 to 2098). In brackets, coefficient of variation (0–1). Source [58].

| Location | Rainfall (%) | ETo (%) | $T_{max}$ (°C) | $T_{min}$ (°C) | $T_{med}$ (°C) |
|---|---|---|---|---|---|
| Archidona | −22.1 (0.2) | 14.5(0.3) | 5.3 (0.38) | 2.3 (0.11) | 3.9 (0.23) |
| Baena | −21.8 (0.2) | 10.8(0.2) | 3.7 (0.20) | 2.1 (0.04) | 3.3 (0.17) |
| Córdoba | −21.0 (0.2) | 9.9(0.2) | 3.6 (0.20) | 2.4 (0.09) | 3.3 (0.18) |
| Mancha Real | −22.0 (0.1) | 15.2(0.4) | 5.4( 0.40) | 2.1 (0.20) | 3.9 (0.24) |
| Average Values | −21.7 | 12.57 | 4.5 | 2.2 | 3.6 |

# 7. CONCLUSIONS

Accelerated soil erosion rates, mostly associated to water erosion are due to a combination of cultivation in sloping areas and to an inappropriate soil management, which has provided limited (or none) ground cover among the olive trees. Although the erosion rates increased with the advent of mechanization of olive cultivation in the early 1960s, our review suggests that they started to be unsustainable since intensification of olive cultivation started in the region during the early 19th century.

Experimental information shows that an appropriate use of cover crops aimed to provide effective ground cover, from fall to early spring, can bring erosion down to tolerable rates, as compared to the effects of bare soil management (CT and NT). Results also indicate an improvement of soil quality when the cover crop soil management is implemented in comparison to CT of NTD.

Erosion model analyses indicate that despite the introduction of cover crop use in the region due to the application of the regulations derived from the EU's Common Agricultural Policy, its impact in achieving tolerable soil erosion rates is still limited. The establishment of effective ground cover, especially in very steep areas, entails a need for additional investments, or difficulties in terms of management operations. In order to achieve tolerable erosion rates there is the need to incorporate a modulation of the requirements to apply ground cover in sloping areas below the 15% threshold currently used, as well as an increase in the extension covered and the growing season, especially in steeper slopes.

The model analysis of two case studies suggests a contrasting response to the impact of decreasing soil depth on olive yield productivity between the sites located in mountainous areas and those located in the rolling landscape of the valleys. Orchards located in valley areas, with soils with relatively high water retention capacity, can stand a significant reduction in soil depth before presenting a marked reduction in yield. This depth, which our analysis estimates around 60 cm, has not been achieved in many of the orchards in that area. The potential productivity of the orchards located in the mountainous areas on soils of limited water holding capacity showed a decrease in yield for any reduction in the soil depth simulated, which are within the range of the soil depth typical of the region. This contrasting behavior has contributed to widening the gap in productivity between both typologies of olive orchards. The hidden cost of the cumulative productivity lost in the mountainous area has been estimated around 200 € $h^{-1}$ $year^{-1}$.

For the more productive orchards, which are the ones located in the valleys, the model analysis also indicates a higher sensitivity to decreasing rainfall as compared to those located in mountainous areas, especially in ranges with low precipitation values (<400 mm). The apparent reason for that is that the current orchard typology in the valleys has been optimized to make the most of infiltrated water. Average annual rainfall values ranging from 400 mm to 500 mm are the expected scenario due to climate change in the region.

The results of this review indicates that more decisive actions should be pursued to curb the high erosion rates in the olive producing areas in the region, with a differentiating motivation and the use of cover crops between the more productive orchards in rolling landscape areas and those located in the mountainous areas.

## ACKNOWLEDGMENTS

This study has been made possible by contributions from different research projects, including P08-AGR-03925 (Andalusian Government), AGL2009-12936-C03-01 (Spanish Ministry of Science and Innovation), AGL2012-40128-C03-01, and AGL2012-40128-C03-02 (Ministry of Economy and Competitiveness) and FEDER funds. Tom Vanwalleghem acknowledges funding by the Spanish Ministry of Economy and Competitiveness (through the Ramón y Cajal Fellowship RYC-2010-07166). We would also like to thank one anonymous reviewer, whose help and careful revision has greatly contributed towards improving the quality of our study.

## AUTHOR CONTRIBUTIONS

Ignacio Lorite contributed in the development and evaluation of the water balance model required for the Section 5 of the manuscript and in Section 6, evaluating the yield function described in this section. Climate scenarios analysis and evaluation of the inputs for the water balance model carried out by him. Encarnación V. Taguas contributed in Section 4 on the evaluation of current erosion rates with experimental data and modelisation analysis. Tom Vanwalleghem contributed inSection 3 analyzing and summarized the data on historical soil erosion. Juan Infante-Amate and Manuel González de Molina contributed on Section 2 on reconstructing the historical evolution of olives, specifically the energy and productive patterns in order to stress the changing functionality and management intensity in the long term. José A. Gómez contributed onSection 3, Section 4, Section 5 and Section 6 on experimental and modelling estimation of erosion rates and effects on soil properties, the development of the water balance model, the correlation between tree transpiration and yield, and its application under diminishing soil depths, as well as leading the overall development of the manuscript. All the authors engaged actively in Section 1 (introduction) and Section 7 (conclusions).

## REFERENCES

1. Beaufoy, G. *EU Policies for Olive Farming. Unsustainable on all Counts*; BirdLife International-WWF: Brussels, Belgium, 2001.

2. Remesal, J. Ben Baso. Revista de la Asociación de Profesores para la Difusión y Protección del Patrimonio. *Escrito en barro: el monte Testaccio y la historia Augusta* **2007**, *16*, 12–15.

3. Zambrana, J.F. *Crisis y modernización del olivar*; Ministerio de Agricultura, Pesca y Alimentación: Madrid, Spain, 1987.

4. Ramón, R.; Williamson, J.G.; Pamuk, I. Specialization in the international market for olive oil before World War I. In*The Mediteranean Response to Globalization before 1950*; Routledge: London, UK, 2000; pp. 159–198.

5. Guzmán, J.R. *El palimpsesto cultivado. Historia de los paisajes del olivar andaluz*; Consejería de Agricultura y Pesca, Junta de Andalucía: Sevilla, Spain, 2004.

6. Infante-Amate, J.; Soto, D.; Cid, A.; Guzmán, G.; González de Molina, M. *Nuevas interpretaciones sobre el papel del olivar en la evolución agraria española. La gran transformación del sector (1880–2010)*; XIV Congreso de la Sociedad Española de Historia Agraria: Badajoz, Spain, 2013.

7. Christiansen, T. Intervención del Estado y mercado negro en el sector oleícola durante el primer franquismo. *Hist. Agrar.* **2002**, *27*, 221–246.

8. Christiansen, T. *The Reason Why The Post Civil-War Agrarian Crisis in Spain*; Prensas Universitarias de Zaragoza: Zaragoza, Spain, 2012.

9. Tió, C. *La política de aceites comestibles en la España del siglo XX*; Ministerio de Agricultura, Pesca y Alimentación: Madrid, Spain, 1982.

10. De Graff, J.; Eppink, L.A.A.J. Olive oil production and soil conservation in southern Spain, in relation to EU subsidy policies. *Land Use Policy* **1999**, *16*, 259–267.

11. Infante-Amate, J. Ecología e historia del olivar andaluz. Un estudio socioambiental de la especialización olivarera en el sur de España (1750–2000). Ph.D. Thesis, Universidad Pablo de Olavide, Sevilla, Spain, 2011.

12. Scheidel, A.; Krausmann, F. Diet, trade and land use: A socio-ecological analysis of the transformation of the olive oil system. *Land Use Policy* **2011**, *28*, 47–56.

13. Infante-Amate, J. The Ecology and History of Mediterranean Olive Grove: The Spanish Great Expansion (1750–2000).*J. Rural Hist.* **2012**, *23*, 161–184.

14. Infante-Amate, J.; González de Molina, M. The Socio-Ecological Transition on a Crop Scale: The Case of Olive Orchards in Southern Spain (1750–2000). *Hum. Ecol.* **2013**, *13*, 961–969.

15. García-Ruiz, R.; González de Molina, M.; Guzmán, G.; Soto, D.; Infante-Amate, J. Guidelines for Constructing Nitrogen, Phosphorus, and Potassium Balances in Historical Agricultural System. *J. Sustain. Agric.* **2012**, *36*, 650–682.

16. Junta Consultiva Agronómica. *Avance estadístico sobre el cultivo y producción del olivo*; Dirección General de Agricultura, Industria y Comercio: Madrid, Spain, 1891.

17. Junta Consultiva Agronómica. *El aceite de oliva*; Dirección General de Agricultura y Montes: Madrid, Spain, 1923.

18. Tombesi, A.; Michelakis, N.; Pastor, M. Recommendations of the Working Group on Olive Farming Production Techniques and Productivity. *Olivae* **1996**, *61*, 38–51.

19. Instituto Nacional de Estadística. Instituto Nacional de Estadística. Censo Agrario 2009. Available online: http://www.ine.es/jaxi/tabla.do (accessed on 25 November 2013).

20. Naredo, J.M. La crisis del olivar como cultivo biológico tradicional. *Agricultura y Sociedad.* **1983**, *26*, 167–277.

21. Ministerio de Agricultura, Alimentación y Medio Ambiente. *Encuesta sobre superficies y rendimientos de cultivos. Análisis de las plantaciones de olivar en España*; Ministerio de Agricultura, Alimentación y Medio Ambiente: Madrid, Spain, 2012.

22. Vanwalleghem, T.; Amate, J.I.; de Molina, M.G.; Fernández, D.S.; Gómez, J.A. Quantifying the effect of historical soil management on soil erosion rates in Mediterranean olive orchards. *Agric. Ecosyst. Environ.* **2011**, *142*, 341–351.

23. Montgomery, D.R. Soil erosion and agricultural sustainability. *Proc. Natl. Acad. Sci.* **2007**, *104*, 13268–13272.

24. González-Hidalgo, J.C.; Peña-Monné, J.L.; de Luis, M. A review of daily soil erosion in Western Mediterranean areas.*Catena* **2007**, *71*, 193–199.

25. Vanwalleghem, T.; Laguna, A.; Giráldez, J.V.; Jiménez-Hornero, F.J. Applying a simple methodology to assess historical soil erosion in olive orchards. *Geomorphology* **2010**, *114*, 294–302.

26. Mabit, L.; Chhem-Kieth, S.; Toloza, A.; Vanwalleghem, T.; Bernard, C.; Amate, J.I.; González de Molina, M.; Gómez, J.A. Radioisotopic and physicochemical background indicators to assess soil degradation affecting olive orchards in southern Spain. *Agric. Ecosyst. Environ.* **2012**, *159*, 70–80.

27. Notebaert, B.; Verstraeten, G.; Rommens, T.; Vanmontfort, B.; Govers, G.; Poesen, J. Establishing a Holocene sediment budget for the river Dijle. *Catena* **2009**, *77*, 150–163.

28. Martín-Puertas, C.; Valero-Garcés, B.L.; Mata, M.P.; Moreno, A.; Giralt, S.; Martínez-Ruiz, F.; Jiménez-Espejo, F. Geochemical processes in a Mediterranean Lake: A high-resolution study of the last 4,000 years in Zoñar Lake, southern Spain. *J. Paleolimnol* **2009**, *46*, 405–421.

29. Gómez, J.A.; Battany, M.; Renschler, C.S.; Fereres, E. Evaluating the impact of soil management on soil loss in olive orchards. *Soil Use Manag.* **2003**, *19*, 127–134.

30. Taguas, E.V.; Giráldez, J.V.; Ayuso, J.L.; Pérez, R.; Castillo, C.; Gómez, J.A. Preliminary assesment and cost analysis of incentives for gully control in agricultural areas of Andalusia. (Southern Spain)Vienna: s.n. *Geophys. Res. Abstr.* **2011**, *13*, 4693.

31. Gómez, J.A.; Giráldez, J.V.; Vanwalleghem, T. Comments on "Is soil erosion in olive groves as bad as often claimed?" by L. Fleskens and L. Stroosnijder. *Geoderma* **2008**, *147*, 93–95.

32. Gómez, J.A.; Giráldez, J.V. Erosión y degradación de suelos. In *Sostenibilidad de la Producción de Olivar en Andalucía*; Consejería de Agricultura y Pesca. Junta de Andalucía: Sevilla, Spain, 2009.

33. Gómez, J.A.; Sobrinho, T.A.; Giráldez, J.V.; Fereres, E. Soil management effects on runoff, erosion and soil properties in an olive grove of Southern Spain. *Soil Tillage Res.* **2009**, *102*, 5–13.

34. Gómez, J.A.; Guzmán, G.; Vanwalleghem, T.; Campos, M.; Giráldez, J.V. Proterra y Biosuelo, siete años de ensayos de cubiertas vegetales para control de la erosión en un olivar de verdeo. *Agricultura* **2010**, *935*, 910–915.

35. Gómez, J.A. Informe de la estación experimental de La Conchuela años hidrológico 2009–2010. In *Internal Report for the RESEL Network of the Spanish Ministry of Agriculture, Food and Environment*; IAS-CSIC: Cordoba, Spain, 2010.

36. Gómez, J.A. Informe de la estación experimental de La Conchuela años hidrológico 2010–2011. Internal Report for the RESEL Network of the Spanish Ministry of Agriculture, Food and Environment, 2011. In *Informe de la Estación Experimental de La Conchuela años Hidrológico 2010–2011. Internal Report for the RESEL Network of the Spanish Ministry of Agriculture, Food and Environment, 2011*; IAS-CSIC: Cordoba, Spain, 2011.

37. Gómez, J.A. Informe de la estación experimental de La Conchuela años hidrológico 2011–2012. In *Internal Report for the RESEL Network of the Spanish Ministry of Agriculture, Food and Environment*; IAS-CSIC: Cordoba, Spain, 2012.

38. Francia, J.R.; Durán, V.H.; Martínez, A. Environmental impact from mountainous olive orchards under different soil-management systems (SE Spain). *Sci. Total Environ.* **2006**, *358*, 46–60.

39. Arroyo, L. Valoración agronómica de las pérdidas por productividad por erosión en cultivos plurianuales mediterráneos. Ph.D. Thesis, Universidad de Cordoba, Cordoba, Spain, 2004.

40. Gómez, J.A.; Vanwalleghem, T.; De Hoces, A.; Taguas, E.V. Hydrological and erosive response of a small catchment under olive cultivation in a vertic soil during a five year period: Implications for sustainability. s.l. *Agric. Ecosyst. Environ.* **2014**, *188*, 229–244.

41. Taguas, E.V.; Ayuso, J.L.; Pérez, R.; Giráldez, J.V.; Gómez, J.A. Intra and inter-annual variability of runoff and sediment yield of an olive micro-catchment with soil protection by natural ground cover in Southern Spain. *Geoderma***2013**, *206*, 49–32.

42. Taguas, E.V.; Burguet, M.; Ayuso, J.L.; Pérez, R.; Gómez, J.A. Interpretation of the impact of different managements and the rainfall variability on the soil erosion in a Mediterranean olive orchard microcatchment. Vienna: s.n.. *Geophys. Res. Abstr.* **2012**, *14*, 10966.

43. Renard, K.G.; Foster, G.R.; Weesies, G.A.; McCool, D.K.; Yoder, D.C. *Predicting Soil Erosion by Water: A Guide to Conservation Planning with the Revised Universal Soil Loss Equation (RUSLE)*; United States Department of Agriculture, Agricultural Handbook. No. 703. Department of Agriculture: Washington, DC, USA, 1997.

44. Marín, V.J. *Interfaz gráfica para la valoración de las pérdida de suelo en parcelas de olivar*; Universidad de Cordoba: Cordoba, Spain, 2013.

45. Instituto para la Conservación de la Naturaleza. *Agresividad de la lluvia en España. Valores del factor R de la ecuación universal de pérdidas de suelo*; Ministerio de Agricultura, Pesca y Alimentación: Madrid, Spain, 1988.

46. Instituto Andaluz de la Reforma Agraria—Consejo Superior de Investigaciones Científicas. *Mapa de suelos de Andalucía (1:400,000)*; Mudarra, J.L., Barahona, E., Baños, C., Iriarte, A., Santos, F., Eds.; Consejería de Agricultura y Pesca, Junta de Andalucía: Sevilla, Spain, 1998.

47. Trueba, C.R.; Millán, T.; Schmid, C.; Roquero, C.; Magister, M. *Bases de datos de propiedades edafológicas de los suelos españoles. Informes Técnicos 882 y 883*; Centro de Investigaciones Energéticas Medioambientales y Tecnológicas: Madrid, Spain, 1999.

48. Guzmán, J.R. *Geografía de los paisajes del olivar andaluz*; Consejería de Agricultura y Pesca. Junta de Andalucía: Sevilla, Spain, 2004.

49. lvarez, S.; Soriano, M.A.; Landa, B.B.; Gómez, J.A. Soil Properties in Organic Olive Orchards as Compared to Natural Areas in a Mountainous Landscape in Southern Spain. *Soil Use Manag.* **2007**, *23*, 404–416.

50. Soriano, M.A.; Álvarez, S.; Landa, B.; Gómez, J.A. Soil properties in organic olive orchards following different weed management in a rolling landscape of Andalusia, Spain. *Renew. Agric. Food Syst.* **2012**, *29*, 83–91.

51. Alcántara, A.M.; Gómez, J.A.; Fereres, E. Evaluación del riesgo de erosión y productividad del olivar en producción integrada. *Agricultura* **2006**, *885*, 426–432.

52. Bennet, H.H. Soil Erosion in Spain. *Geogr. Rev.* **1960**, *50*, 59–72.

53. Milgroom, J.; Soriano, M.A.; Garrido, J.M.; Gómez, J.A.; Fereres, E. The influence of a shift from conventional to organic olive farming on soil management and erosion risk in southern Spain. *Renew. Agric. Food Syst.* **2007**, *22*, 1–10.

54. Moriana, A.; Orgaz, F.; Pastor, M.; Fereres, E. Yield responses of a mature olive orchard to water déficits. *J. Am. Soc. Hort. Sci.* **2003**, *128*, 425–431.

55. Abazi, U.; Lorite, I.J.; Cárceles, B.; Martínez Raya, A.; Durán, V.H.; Francia, J.R.; Gómez, J.A. WABOL: A conceptual water balance model for analyzing rainfall water use in olive orchards under different soil and cover crop management strategies. *Comput. Electron. Agric.* **2013**, *91*, 35–48.

56. Mesa-Jurado, M.A.; Berbel, J.; Orgaz, F. Estimating marginal value of water for irrigated olive grove with the production function method. *Span. J. Agric. Res.* **2010**, *8*, 197–206.

57. Lal, R. Soil erosion impact on agronomic productivity and environmental quality. *Crit. Rev. Plant Sci.* **1998**, *17*, 319–464.

58. Rodríguez, M.T.; Lorite, I.; Ruiz, M.; Dosio, A.; Gómez, J.A. Impact of climate change on water balance components in Mediterranean rainfed olive orchards under tillage or cover crop soil management. Vienna. s.n.. *Geophys. Res. Abstr.***2013**, *15*, 11446.

# Chapter 8

# Alternative Land Management Strategies and Their Impact on Soil Conservation

**Tiziano Gomiero**

ICTA, Universitat Autonoma de Barcelona, Campus de Bellaterra, 08193 Cerdanyola del Valles, Spain

## ABSTRACT

Soil conservation is threatened by a number of factors, namely the effects of intensive agricultural practices, the increasing pressure for food production linked to the increasing human population, the consumption patterns in developed and emerging economies, and the conversion of agriculture from the production of commodities (which is itself a goal in need of discussion) to the production of biofuels. The extent of human pressure and the effects of conflicting land use systems need to be addressed. Alternative and conservative agricultural practices need to be explored and widely adopted in order to preserve the soil fertility, assessing their pros and cons. In this paper, the main potential alternative practices are reviewed, focusing in particular on organic farming. It is also argued that in order to better plan to preserve soil health a strategy considering the whole food system is required.

## KEYWORDS

soil conservation practices; organic farming; low-input farming; land use change; farming system analysis


## 1. INTRODUCTION

In the early 1970s, with the "green revolution" led by Norman Borlaug (1914–2009, Nobel Peace Prize in 1970), agricultural productivity experienced an incredible leap forward. The productivity of the new high-yielding grain varieties (HYVs, termed also high-response varieties), on average, more than doubled and for some cereals it improved by a staggering 4–5 times [1, 2, 3]. This helped to meet world food demand and saving hundreds of millions from starvation. Asia, for example, which was threatened by hunger and mass starvation as late as the mid-1960s, became self-sufficient in staple foods within 20 years, even though its population more than doubled [4]. The green revolution relied on cheap fossil fuels, which allowed the industrial production of chemical fertilizers and pesticides, the mechanization of agriculture and the massive use of irrigation [1,5]. Along with those tangible benefits, the new intensive agriculture had to face also some problems. Crops were more prone to pests and diseases and newly developed synthetic pesticides were necessary to keep pests out of the crops. The environmental impact of agricultural activity increased too, and the overall efficiency (as output/input) declined sharply [2,3,4,6,7]. Decades of agricultural intensification, then, exacerbated the demand for water use and affected soil fertility, threatening long-term crop productivity by increasing soil degradation and causing water shortages [7,8,9,10].

It has been widely reported that intensive farming exacerbates soil erosion and loss of Soil Organic Matter (SOM), threatening the long-term sustainability of crop production, especially under extreme climatic events such as droughts [9,11,12,13,14,15,16].

About 40% of global croplands may be experiencing some degree of soil erosion, reduced fertility, or overgrazing [9,11,17,18]. Soil erosion has been estimated to reduce yields on about 16% of agricultural land, especially cropland in Africa and Central America and pastures in Africa [17]. At present, the accelerated rates of erosion experienced are causing major modifications to carbon, nitrogen and phosphorus biogeochemical cycles

[10,19]. Resistance of soils to erosion is closely linked to the stabilizing influence of organic matter and vegetation cover. In regions such as Asia and Africa, where soil erosion is associated with reduced vegetation cover, the loss of soil carbon can trigger catastrophic shifts to severely degraded landscapes [10,20].

Most SOM is found in the topsoil (15–25 cm of the A horizon) in the form of above- and belowground plant residues. SOM is of key importance for soil fertility [11,14,21,22,23,24]. The Soil Organic Carbon pool to 1 m depth ranges from 30 tons $ha^{-1}$in arid climates to 800 tons $ha^{-1}$ in organic soils in cold regions, and a predominant range of 50 to 150 tons $ha^{-1}$ [14]. Fertile agricultural soils can contain up to 100 tons of organic matter per hectare (or 4% of the total soil weight) and, in the case of most agricultural soils, SOM represents 1%–5% of topsoil [25]. Conventional agricultural practices that tend to leave soil uncovered for long periods of the year are responsible for topsoil erosion and reduction of its SOM content. Soil removed by either wind or water erosion is 1.3–5.0 times richer in organic matter than the soil left behind [14,15,21,26]. About 95% of soil nitrogen and 25%–50% of soil phosphorus are contained in the SOM-containing topsoil layer [15,21], the importance of which is such that in one study it was estimated that the reduction of SOM from 1.4% to 0.9% lowered the grain yield potential by 50% [27].

Agricultural practices such as no-till agriculture or minimum tillage, and organic farming can help to reduce soil loss and restore soil fertility [15,16,28,29,30,31]. Although permanent grasslands respond slower and probably weaker to organic management than crop fields do, it seems that organic practice may benefits grassland too [32].

In this paper, I will review some key issues concerning the environmental impact of current agricultural practices and present some potential alternative agricultural practices that may be adopted to make agriculture less environmentally damaging, reducing the use of natural resources and preserving soil fertility and biodiversity. Finally, I will discuss the need to adopt a complex approach when planning for land use and soil management.

## 2. SOIL IN CONTEXT: SOCIOECONOMIC PRESSURE AND IMPACT OF CURRENT AGRICULTURAL PRACTICES

Agriculture should also aim at guarantying food security for people. As stated by [33], food security is defined as a state when *"all people, at all times, have physical and economic access to sufficient, safe and nutritious food for a healthy and active life"*.

Agricultural systems are agroecosystems and agricultural science can be referred to as agroecology [22,34,35]. To this regard Miguel Altieri, one of the pioneer in agroecology science [22,36] provides the following definitions: "***Agroecosystems****are communities of plants and animals interacting with their physical and chemical environments that have been modified by people to produce food, fiber, fuel and other products for human consumption and processing.* ***Agroecology*** *is the holistic study of agroecosystems including all the environmental and human elements. It focuses on the form, dynamics and functions of their interrelationship and the processes in which they are involved.*" (p. 8 in [36]).

Agroecosystems can effectively be understood as made up of many different component,s which can only be defined after adopting different perceptions/descriptions referring to different levels and scales. When adopting a *biophysical view* we can move from the scale of the soil, and its fauna, to arrive at the scale of crop species. When adopting a *socioeconomic view* we can start from the scale of the farmer, and moving up to that of the household of the rural community in which the household lives, to arrive at the scale of local, regional and national socio-economic systems. Using an *ecological view* we can have: the scale of the micro-ecosystems in the soil, local agroecosystems, then the scale of landscape and the watershed, up to the natural mechanisms generating biogeochemical cycles of water and nutrients that are required to stabilize and constrain the boundary conditions of the farming activity [37,38]. Agriculture is dynamic and evolving in the sense that it reflects both the changes occurring in: (i) external constraints, posed by the environmental system (e.g., biodiversity, landscape, and climate) and the socioeconomic systems (e.g., demography, cost of labor, and technological innovation), and (ii) internal characteristics reflecting cultural, social and technical innovations (Figure 1).

However, this complex nature of agroecosystem has been neglected. As pointed out by Foley *et al.* (pp. 570–571 in [39]) "In short, modern agricultural land use practices may be trading short-term increases in food production for long-term losses, in ecosystem services, including many that are important to agriculture". In order to work for a more sustainable agriculture we have to acknowledge: (1) the multifunctional nature of agriculture, that is not only producing commodities but also preserving the health of ecosystems, and quality of life for consumers and rural communities; and (2) the multi-scale nature of the complex network of relations among ecosystems and socioeconomic systems, that requires considering simultaneously different but relevant dynamics operating at different hierarchical levels [37,38,39,40,41,42].

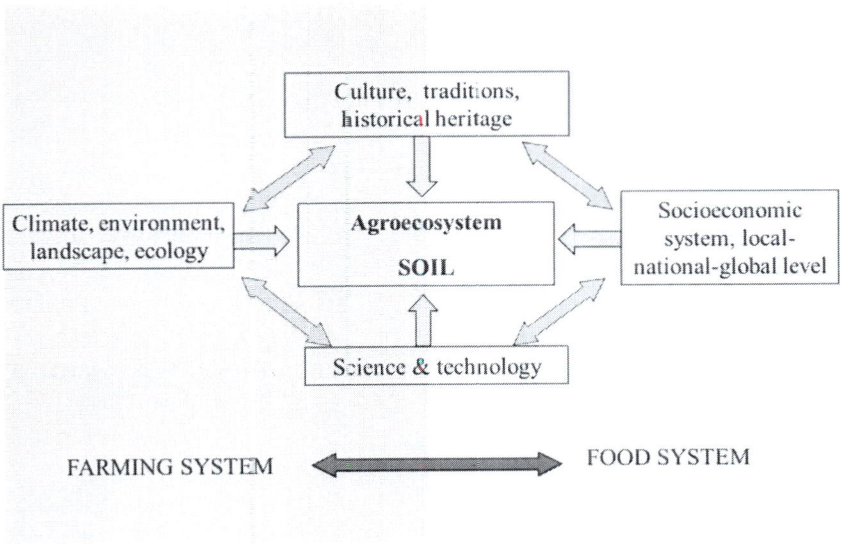


**Figure 1.** Soil and agroecosystems are complexly interwoven with the environmental, ecological and socioeconomic domains, at different hierarchical scales.

Soil is the keystone of the agroecosystem, and preserving its health and fertility is a must. In order to work toward better soil management, many different issues must be tracked and dealt with: from adopting healthier management practices, to dealing with socioeconomic problems (e.g., population pressure, income distribution, market structure, subsidies mechanism), up to changing some of our paradigms and "values" concerning what agriculture is for (e.g., commodity production *vs.* food security) and how it should be cared for (e.g., massive exploitation *vs.* long term preservation of our "living support system"). Some key challenges ahead concern, for example:

***(1) Population****:* Human population is rising by about 75 million (1.1%) per year, and will reach the staggering figures of 8.3 billion by the 2030 and of 9.2 billion by the 2050 [43,44,45]. At the same time (i) crop productivity stagnates and the yearly supply of grain per capita is decreasing [46,47,48], (ii) about 2 billion ha of the world's agricultural land have been degraded [9,23,39,47], and (iii) the Human Appropriation of Net Primary Productivity (HANPP) reached 50% leaving less and less room and resources to biodiversity and ecosystems, thus compromising the existence of many species and the proper functioning of ecosystems [49,50,51].

***(2) Land***: Recent studies suggest that by 2050 the world will need 60% [45], up to 70 more food [52]. By 2030, worldwide, an additional 120 million ha—an area twice the size of France—will be needed to support food production [53], while built-up area/cropland area—estimated 3.5% in 2000—will reach 5.1% (then 7% in 2050) [47]. The Millennium Ecosystem Assessment [7] estimated that, roughly, 10%–20% (low to medium certainty) of current grassland and forestland is projected to be converted to other uses between now and 2050, mainly due to the expansion of agriculture and, second, due to the expansion of cities and infrastructure. So a new challenge lies ahead—how to feed 9 billion with less land, water and energy [54,55]?

***(3) Land for food* vs. *land for energy***: Biofuels have lately been indicated as a promising source of cheap and sustainable energy. However, large scale conversion of crops, grasslands, natural and semi-natural ecosystems may have detrimental social (e.g., food security, land, grabbing) and ecological consequences (e.g. deforestation, loss of biodiversity, soil erosion) [56,57]. Further to that the power density ($W/m^2$) of biomass is simply too low to fuel our energivorous society [5,55,58]. Such low power density makes biofuels a highly labor-intensive energy source, resulting in the displacement of a high fraction of society labor force to energy production and so an increase of the price of energy [5,55,59].

***(4) Water***: Currently, on a global scale, 70% of the 3800 $km^3$ of water that humans use is directed towards agriculture, 20% towards industry and 10% towards urbanized areas [8]. By 2050, agricultural water use is expected to increase by 13% [8]. The production of common crops in many parts of the world requires a great amount of water, from a few hundreds to a few thousand times the final crop mass [1,3,8,60,61].

***(5) Biodiversity***: Agricultural expansion has a direct impact on local biodiversity through landscape modification which, in turn, results in displacement of local populations and loss of ecosystem services. The loss of native habitats and agricultural intensification, which displaces traditional varieties of seeds with modern high-yielding, but genetically uniform crops, are threatening biodiversity (both wild and domesticated) all over the globe [7,39,62,63,64,65,66]. Farming, including land conversion to farmland, for instance, accounts for 37% of threats to bird species listed as threatened species [65]. Extensive industrialized agriculture also greatly contributes to impoverishing crop biodiversity, with the loss of a large number of agricultural species and varieties [67].

***(6) Synthetic fertilizers***: Between 1960 and 1995, at a global scale, N fertilizer use on cereals increased sevenfold, whilst cereal yields more than doubled; however, N fertilizer efficiency (cereal yields divided by N fertilizer inputs) declined from over 70 to around 25 kg grain per kg N [6,68]. The overall global nitrogen use efficiency of cereals decreased from ~80% in 1960 to ~30% in 2000 [6,69]. Global data for maize, rice, and wheat indicate that only 18% to 49% of nitrogen applied as fertilizer is taken up by crops; the remainder is lost to runoff, leaching, or volatilization [68]. Nitrogen fertilizers are of key importance in intensive conventional agriculture. However, their use turns out to be a major cause of concern when it comes to environmental pollution. The primary source of N pollution comes from N-based agricultural fertilizers, whose use is forecast to double or almost triple by 2050 [6,19,70]. Synthetic fertilizers also have a direct negative impact on soil diversity and several soil processes. The use of synthetic fertilizers and herbicides applications changes relations within and between below and aboveground soil components, promoting negative impact on agriculture by reducing internal biological cycles and pest control [71].

***(7) Pesticides***: Widespread use of pesticides on crops has led to the emergence of many pesticide-resistant pests and pathogens [62,72,73,74,75]. People can be exposed to excessive pesticide levels while working; via food, soil, water or air; or by directly ingesting pesticide products. Pesticides are known to cause 26 million human poisonings per year and 220,000 deaths [76]. Along with other synthetic chemicals, some pesticides have a direct effect on the reproductive system of many high organisms, acting as endocrine disruptors, and inducing severe reproductive problems and modifying sexual behavior [77,78].

***(8) Food waste***: When it comes to the food system, it is disturbing, to say the least, to know that 30%–40% of the food produced in the field is wasted through the food system; in industrialized countries it is estimated that 15%–20% of it just passes directly from our fridge to the bin [1,54,79,80,81,82]. Food wastage clashes with the increasing costs of intensive agricultural practices. Costs are being paid in terms of loss of soil and fertility, reduction of water supply, threat to biodiversity and pollution from agrochemicals [6,7,8,19]. For the USA it has been estimated that the energy embedded in wasted food represents approximately 2% of annual energy consumption, which is substantial when compared to other energy conservation and production proposals [80].

***(9) Diet balance***: Population growth is not the only driver increasing the demand for agricultural products. Higher pressure is exerted by the increasing per capita consumption and changes in diets leading to the consumption of more livestock products by people in newly industrialized countries such as China, India and Brazil [3,45,83]. Referring to the average American diet with meat, Pimentel and Pimentel [3], and Pimentel [84] estimate that a vegetarian diet of an equivalent 2200 kcal per day (referring to The Food and Drug Administration, which recommends an average daily consumption of 2000 kcal for females and 2500 kcal per day for males), requires 33% less fossil energy, much less than the average American is consuming today. Reducing the caloric intake would significantly reduce the total energy expended for food production as well as help lessen the obesity problem [3,83].

***(10) Energy***: The global share of energy used in agriculture is less than 5% of all primary inputs, ranging in affluent societies from as low as 3%, in the United States, and as high as 11% in the Netherlands [5]. In contrast, direct and indirect agricultural energy uses claimed about 15% of China's total primary energy supply [5]. About 30%–50% of agriculture energy cost is accounted for fertilizers [3,5,82]. In industrialized countries, however, the energy in the production sector (agriculture) is a small fraction of the total energy cost of food, as most of the energy used by the food system being consumed by post-harvest activities (Table 1). In order to save energy and reduce $CO_2$ emission related to food production and consumption, it is then important to address the functioning of the whole food system. Pimentel *et al.* [85] estimated that the amount of fossil energy used in the USA food system could be reduced by about 50% with changes in production, processing, packaging, transport, and consumption.

**Table 1.** Energy input for a 455 g, can of sweet corn in the USA market (from [3] modified).

| Energy input | Kcal | % On total |
|---|---|---|
| Production | 450 | 15 |
| Processing | 316 | 10 |
| Packaging | 1006 | 33 |
| Transport | 158 | 5 |
| Distribution | 340 | 11 |
| Shopping | 311 | 10 |
| Home | 457 | 15 |
| Total | 3038 | 100 |

# 3. ALTERNATIVE MANAGEMENT PRACTICES

In this section, some promising alternative agricultural practices are reviewed concerning their benefits to preserve soil fertility and reduce the overall environmental impact of agriculture.

## 3.1. Low Input, No-Tillage Agriculture

Sustainable agriculture must, as defined by the U.S. Department of Agriculture in the 1990 Farm Bill: "over the long term, satisfy human needs, enhance environmental quality and natural resource base, make the most efficient use of nonrenewable resources and integrate natural biological processes, sustain economic viability, and enhance quality of life" [86]. The early idea of a sustainable agriculture was for a farming system that mimics natural ecosystems, and developed in USA in the 1980s (e.g., [87,88,89]).

Sustainable agriculture should aim at preserving the natural resource base, especially soil and water, relying on minimum artificial inputs from outside the farm system. It should be able to recover from the disturbances caused by cultivation and harvest while at the same time being economically and socially viable [22,34,35,88,90,91,92,93,94,95,96]. Sustainable agriculture aims at keeping to a minimum or not used at all agrochemicals (such as in organic agriculture), by adopting conservative practices (crop rotation, integrated pest management, natural fertilization methods, minimum tillage, biological control), which are fully integrated in farm management. Agricultural practices such as no-till agriculture, or minimum tillage, have demonstrated valuable strategy to reduce soil loss and restore soil fertility [16,28,36,91,92]. Soil fertility can be improved also in conventional tillage, by adopting proper farming practices (e.g., using cover crops, leaving residues in the field, avoiding soil compaction, reducing the use of agrochemicals).

Agricultural activities (not including forest conversion) account for approximately 5% of anthropogenic emissions of $CO_2$and the 10%–12% of total global anthropogenic emissions of GHGs (5.1 to 6.1 Gt $CO_2$ eq. $yr^{-1}$ in 2005), accounting for nearly all the anthropogenic methane and one to two thirds of all anthropogenic nitrous oxide emissions are due to agricultural activities [97]. In 2008, in the USA, agricultural activities were responsible for about 7% of total U.S. GHGs emissions in 2008 (with livestock as major contributors) with an increase of 10% from 1998 to 2008 [98]. Adopting conservation tillage (including no-till practices), could reduce GHGs by an important amount [14,15,99,100,101]. It has been estimated [102] that by adopting sustainable practices, agriculture could offset up to about 20% of total global annual $CO_2$ emissions. Evidence from numerous Long Term Agroecosystem Experiments indicates that returning residues to soil, rather than removing them, converts many soils from “sources” to “sinks” for atmospheric $CO_2$ [12,14,101,102]. According to [101], it seems possible that a mix of residues that includes legumes and more recalcitrant materials such as cereal straw or forage crops could result in the greatest stabilization of soil structure.

However, it is to be pointed out that SOM can be increased up to a certain level where it starts leveling-off. Therefore, there is a limit to how much carbon the soil can capture acting as a carbon sink and that conversion to more sustainable agriculture can only represent a temporary and partial solution to the problem of carbon dioxide emissions. Long term solutions concerning GHGs emission abatement should rely on a more general change of our development path, for instance by reducing the overall fossil fuel consumption.

## 3.2. Perennial Crops

From the 1980s some scholars began suggesting to move from an agriculture based on annual crops to an agriculture relying on the cultivation of perennial crops, order to reduce the detrimental effect of soil tillage; agrochemical usage could be avoided or at least greatly reduced authors. Perennial crops seem to bring a valuable number of benefits [68,87,88,103,104]: (i) with their roots exceeding depths of two meters, can improve ecosystem functions, such as water conservation, nitrogen cycling and carbon sequestration; more than 50% when compared to conventional crops, they are reported to be 50 times more effective than annual crops in maintaining topsoil, reduce N losses from 30 to 50 times, and store about 300 up to 1100 kg C/ha per year compare to 0 to 300–400 kg C/ha per year as do annual crops, and it is believed they could help restrain climate change, (ii) reduce management costs, because they do not need to be replanted every year; so they require fewer passes of farm machinery and fewer inputs of pesticides and fertilizers as well, which reduces fossil fuel use, (iii) required less harmful inputs, herbicide costs for annual crop production may be 4 to 8.5 times the herbicide costs for perennial crop production, so fewer inputs in perennial systems mean lower cash expenditures for the farmer and a greatly reduced environmental impact. Work is still needed, however, to improve yield performances and post-harvest processing technologies capable of utilizing more perennial crops [103,104].

Perennial crops are predicted to better adapt to temperature increases as the magnitude predicted by most climate-change models. Cassman *et al.* [68] report that increases of 3 to 8 degrees Celsius are predicted to increase yields of switchgrass (*Panicum virgatum*), a perennial forage (used as hay or grazed by cattle) and energy crop, by 5000 kg per ha, whereas for annual species yields are predicted to decline (e.g., maize—1500 kg per ha; soybean—800 kg per ha; sorghum—1000 kg per ha).

## 3.3. Organic Farming

A different alternative to conventional agriculture has been proposed and implemented by the organic agriculture movement. Organic agriculture is regulated by international and national institutional bodies, which certify organic products from production to handling and processing [105,106,107,108,109,110]. The origins of organic agriculture can be traced back to the 1920–1930 period in North Europe (mostly Germany and UK) [30,111,112,113,114], and it is now widely spread all over the world.

Preserving and enhancing soil health is at the core of organic farming. To preserve soil fertility organic agriculture relies on a number of farming practices that take fully advantage of the ecological cycles. In organic farming systems, soil fertility is enhanced by crop rotation, intercropping, polyculture, cover crops and mulching. Pest control is achieved by using appropriate cropping techniques, biological control and natural pesticides (mainly extracted from plants). Weed control, in many cases the main focal problem for organic farming, is managed by appropriate rotation, seeding timing, mechanic cultivation, mulching, transplanting, flaming, *etc.*, [15,73,112,115].

Soil characteristics are generally site-specific, but to date many studies have proven organic farming to perform better in preserving or improving soil quality with regards to both biophysical (e.g., SOM, stored nutrients) and biological (e.g., biodiversity) properties (e.g., [30,99,116,117, 118,119, 120, 121, 122,123,124,125,126,127,128,129,130]). It has also being reported that the combined application of organic resources and mineral fertilizers may represent a viable approach to address soil fertility decline in sub-Saharan Africa [131].

Although few in number, important long-term studies concerning SOM content and soil characteristics in organic and conventional soils have been carried out, both in USA and Europe [30,116,121,124,129]. Organic farming resulted in significant higher SOM content along with other better biochemical performance indicators and soil erosion was greatly reduced, with an erosion rate that differed up to 4 times [124].

Techniques to reduce N loss and to increase the efficiency of N uptake are widely used in organic farming, and many trials demonstrate the benefit of organic farming in reducing N leaching and increasing N uptake efficiency [99,117,132,133]. Possible drawbacks from organic fertilization have been reported by some authors (e.g., [2,134,135,136]): the "slow release" of nutrients from organic compost or green manures can be difficult to control and harness and may fail to match crop demand, resulting in N losses through leaching and volatilization. Moreover, in organic systems, competition with weeds can greatly reduce N intake efficiency [135].

Sustainable agricultural practices can be effective in improving water use efficiency in particular in poor developing country affected by water scarcity [93,92]. Long-term crop yield stability and the ability to buffer yields through climatic adversity will be critical factors in agriculture's capability to support society in the future. It has been estimated that for every 1% of SOM content, the soil can hold 10.000–11.000 liters of plant-available water per ha of soil down to about 30 cm [13]. A number of studies have shown that, under drought conditions, crops in organically managed systems produce higher yields than comparable crops managed conventionally. This advantage can result in organic crops out-yielding conventional crops by 70%–90% under severe drought conditions [122,124,137,138,139]. The primary reason for higher yield in organic crops is thought to be due to the higher water-holding capacity of the soils under organic management, up to 100% higher in the crop root zone [13,116,122]. Others studies have shown that organically managed crop systems have lower long-term yield variability and higher cropping system stability [12,138].

A number of works [30,123,140,141], reports that organic farming is usually associated with a significantly higher level of biological activity, represented by bacteria, fungi, springtails, mites and earthworms. In a Swiss long-term experiment [119,121,127], soil ecological performance was greatly enhanced under biodynamic and organic management. Microbial biomass and activity increased under organic management, root length colonized by mycorrhizae in organic farming systems was 40% higher than in conventional systems, and biomass and abundance of earthworms were from 30% to 320% higher in the organic plots as compared with conventional. Concerning soil health, Briar *et al.* [126] conclude that transition from conventional to organic farming can increase soil microbial biomass and populations of beneficial bacterivore nematodes while simultaneously reducing the populations of predominantly plant-parasitic nematodes. Soil management is reported to affect pest response. A number of studies report pest preferring plants which have been nurtured with synthetic fertilizer rather than those growing in organically managed soil [30,142,143,144,145,146,147], and under conventional management natural enemies are also often doing worse probably due to the use of agrochemicals [148].

However, it has to be pointed out that local specificity plays an important role in determining the performance of a farming system (and hold true for management systems in general): what is sustainable for one region may not be for another region or area [138,149]. This is relevant to understand both the unifying principles of soil response over large scale [150], and across different land-use types [71]. For example, in highly densely populated areas of Asia it may be difficult to give up chemical fertilizers without posing a risk to food security. Still sound techniques could be employed to limit the use of harmful (and costly) pesticides, prevent soil erosion and protecting biodiversity [36,73,151,152,153,154].

The benefits associated with the adoption of organic farming practices have been questioned by many authors to different degrees; I refer to Gomiero *et al.* [30] for a review of the discussion. Although yield in organic agriculture is generally lower (20% or more) than that from conventional farming [30,155,156] (but there are long term studies which prove that the yield can be comparable (e.g., [124]), still it is important to conduct more research in that direction in view of the foreseen higher cost of agricultural inputs (along with the increasing cost of energy), and the need to preserve the fertility of the soil. It has also to be pointed out that the assessment of agriculture cannot be limited to yield or commodities production, or account only for farm investment and revenue [30,149]. Other key issues have to be accounted for, which are generally missed such as: water consumption, biodiversity preservation, reduction in the use of agrochemicals, energy consumption and GHGs emission.

## 4. CONCLUSIONS

Intensive farming tends to exacerbate soil erosion and loss of soil organic matter, posing a threat to the long term sustainability of agriculture, especially under extreme climatic events such as droughts.

Work is urgently needed to improve the overall efficiency of the agricultural and food system, in a way that reduces the pressure exerted on the environment and on the soil. For that we have to simultaneously deal with a number of different and complex problems of widely different nature: environmental, social, economic, technical, cultural, educational, *etc.* We cannot think to save our soil if we keep treating it as a factory, sticking to a "fuel in–food out" model. Soil is not a machine, it is a living being of astonishing, and still largely unexplored, complexity. To seriously address the soil issue, we should agree that it is necessary to rethink the structure and functioning of the whole food system to start with. Social policies (e.g., education, women empowerment, microcredit, participative extension and health programs) should be undertaken in order to improve the livelihood conditions of those affected by poverty and high demographic growth.

Farming techniques able to effectively reduce soil erosion, while preserving and enhancing fertility, have been developed and are adopted by farmers in many different regions. Reduced tillage and organic farming, along with a shift to perennial crops are providing promising results and should be further explored and researched, particularly concerning weeds and pest control. To date many studies have proved organic farming to perform better than conventional in improving soil quality regarding to both biophysical (e.g., SOM, stored nutrients) and biological (e.g., biodiversity) properties. Critics may be right in pointing out some inherent limits of organic agriculture (e.g., lower yield). The sustainability of the food system, however, has to be addressed from many different perspectives, in a holistic way, and with a long term perspective in mind. Given the crucial role that food production (and then agriculture) plays in our life, our major concern should be to secure that farming practice guarantees the resilience of our food production system (*"The ability of a system to maintain productivity in spite of a major disturbance, such as caused by intensive stress or a large perturbation"* [157], p. 723). More work has to be done to acquire knowledge about the comparative sustainability of different farming systems. Adaptive measures to cope with climate change should treasure knowledge gained from organic and other alternative farming practices. Extensive experimentation should be conducted to gain better understanding of the complex interaction among farming practices, environmental characteristics and agroecosystem resilience.

Eventually we should be willing to reconsider our relation with the land and rethink the structure of our food system in view of increasing its sustainability. That implies to work in parallel also on the social, economic and political dimensions of our society. I wish to conclude with the words of Aldo Leopold (p. 226 in [158]) *"Perhaps the most serious obstacle impeding the evolution of a land ethic is the fact that our educational and economic system is headed away from, rather than toward, an intense consciousness of land"*. It is time to go back to the foundation.

## ACKNOWLEDGMENTS

I wish to thank four anonymous reviewers for their valuable comments. I also wish to acknowledge David Pimentel (Cornell University) for his invitation to contribute to this special issue and for his comments on the first draft of the paper.

## REFERENCES


1. Smil, V. *Feeding the World: A Challenge for the Twenty-First Century*; The MIT Press: Cambridge, MA, USA, 2000.

2. Tilman, D.; Cassman, K.G.; Matson, P.A.; Naylor, R.; Polasky, S. Agricultural sustainability and intensive production practices. *Nature* **2002**, *418*, 671–677.

3. Pimentel, D.; Pimentel, M. *Food, Energy, and Society*, 3rd ed.; CRC Press: Boca Raton, FL, USA, 2008.

4. Hazell, H.; Wood, S. Drivers of change in global agriculture. *Phil. Trans. B* **2008**, *363*, 495–515.

5. Smil, V. *Energy at the Crossroads*; The MIT Press: Cambridge, MA, USA, 2003.

6. Tilman, D.; Fargione, J.; Wolff, B.; D'Antonio, C.; Dobson, A.; Howarth, R.; Schindler, D.; Schlesinger, W.H.; Simberloff, D.; Swackhamer, D. Forecasting agriculturally driven global environmental change. *Science* **2001**, *292*, 281–284.

7. Millenium Ecosystem Assessment. Ecosystems and Human Well-Being: Biodiversity Synthesis. World Resources Institute: Washington, DC, USA, 2005. Available online: http://www. millenniumassessment. org/documents/ document.354.aspx.pdf (accessed on 20 July 2009).

8. Molden, D., Ed.; *Water for Food, Water for Life. A Comprehensive Assessment of Water Management in Agriculture*; Earthscan: London, UK, 2007. Available online: http://www.iwmi. cgiar.org/assessment /Publications/books.htm(accessed on 10 May 2013).

9. Montgomery, D.R. Soil erosion and agricultural sustainability. *Proc. Natl. Acad. Sci. USA* **2007**, *104*, 13268–13272.

10. Quinton, J.N.; Govers, G.; van Oost, C.; Bardgett, R.D. The impact of agricultural soil erosion on biogeochemical cycling. *Nat. Geosci.* **2010**, *3*, 311–314.

11. Pimentel, D.; Harvey, C.; Resosudarmo, P.; Sinclair, K.; Kurz, D.; McNair, M.; Crist, S.; Sphpritz, L.; Fitton, L.; Saffouri, R.; *et al*. Environmental and

economic costs of soil erosion and conservation benefits. *Science* **1995**, *267*, 1117–1123.

12. Rasmussen, P.E.; Goulding, K.W.T.; Brown, J.R.; Grace, P.R.; Janzen, H.H.; Körschens, M. Long-term agroecosystem experiments: Assessing agricultural sustainability and global change. *Science* **1998**, *282*, 893–896.

13. Sullivan, P. *Drought Resistant Soil*; ATTRA, National Center for Appropriate Technology USDA: Washington, DC, USA, 2002. Available online: http://attra.ncat.org/attra-pub/PDF/drought.pdf (accessed on 24 November 2009).

14. Lal, R. Soil carbon sequestration impact on global climate and food security. *Science* **2004**, *304*, 1623–1627.

15. Lal, R. Enhancing eco-efficiency in agro-ecosystems through soil carbon sequestration. *Crop Science* **2010**, *50*, 120–131.

16. NRC (National Research Council). *Toward Sustainable Agricultural Systems in the 21st Century*; National Academies Press: Washington, DC, USA, 2010. Available online: http://www.nap.edu /catalog/ 12832. html (accessed on 3 March 2012).

17. Wood, R.; Lenzen, M.; Dey, C.; Lundie, S. A comparative study of some environmental impacts of conventional and organic farming in Australia. *Agric. Syst.* **2006**, *89*, 324–348.

18. Reynolds, J.F.; Stafford Smith, D.M.; Lambin, E.F.; Turner, B.L., II; Mortimore, M.; Batterbury, S.P.J.; Downing, T.E.; Dowlatabadi, H.; Fernández, R.J.; Herrick, J.E.; *et al*. Global desertification: Building a science for dryland development. *Science* **2007**, *316*, 847–851.

19. Vitousek, P.M.; Naylor, R.; Crews, T.; David, M.B.; Drinkwater, L.E.; Holland, E.; Johnes, P.J.; Katzenberger, J.; Martinelli, L.A.; Matson, P.A.; *et al*. Nutrient imbalances in agricultural development. *Science* **2009**, *324*, 1519–1520.

20. Berhe, A.; Harte, J.; Harder, J.; Torn, M. The significance of the erosion-induced terrestrial carbon sink. *BioScience***2007**, *57*, 337–346.

21. Allison, F.E. *Soil Organic Matter and Its Role in Crop Production*; Elsevier: Amsterdam, The Netherlands, 1973.

22. Altieri, M.A. *Agroecology: The Science of Sustainable Agriculture*; Westview Press: Boulder, CO, USA, 1987.

23. Pimentel, D.; Kounang, N. Ecology of soil erosion in ecosystems. *Ecosystems* **1998**, *1*, 416–426.

24. Bot, A. *The Importance of Soil Organic Matter: Key to Drought-Resistant Soil and Sustained Food Production*; FAO: Rome, Italy, 2005. Availableonline: http://www.fao.org/docrep/009/a0100e/a0100e00.htm#Contents (accessed on 20 June 2009).

25. Russell, E.W. The role of organic matter in soil fertility. *Phil. Trans. B* **1977**, *281*, 209–219.

26. Barrows, H.L.; Kilmer, V.J. Plant nutrient losses from soils by water erosion. *Adv. Agron.* **1963**, *15*, 303–315.

27. Liebert, B. *The Environmental Heritage of Soviet Agriculture*; CABI: Oxon, UK, 1995.

28. Lal, R. Evolution of the plow over 10,000 years and the rationale for no-till farming. *Soil Tillage Res.* **2007**, *93*, 1–12.

29. Gomiero, T.; Pimentel, D.; Paoletti, M.G. Is there a need for a more sustainable agriculture? *Crit. Rev. Plant. Sci.* **2011**,*30*, 6–23.

30. Gomiero, T.; Pimentel, D.; Paoletti, M.G. Environmental impact of different agricultural management practices: Conventional *vs.* organic agriculture. *Crit. Rev. Plant. Sci.* **2011**, *30*, 95–124.

31. Carr, P.M.; Gramig, G.G.; Liebig, M.A. Impact of organic zero tillage systems on crops, weeds and soil quality.*Sustainability* **2013**, *5*, 3172–3201.

32. Klaus, V.H.; Kleinebeckera, T.; Prati, D.; Gossner, M.M.; Alt, F.; Boch, S.; Gockel, S.; Hemp, A.; Lange, M.; Müller, J.; *et al.* Does organic grassland farming benefit plant and arthropod diversity at the expense of yield and soil fertility? *Agric. Ecosys. Env.* **2013**, *177*, 1–9.

33. World Food Summit 1996. *Declaration on World Food Security*; FAO: Rome, Italy. Available online: http://www.fao. org/docrep/003/ w3 613e/ w3613e00.htm (accessed on 4 August 2013).

34. Wezel, A.; Soldat, V. A quantitative and qualitative historical analysis of the scientific discipline of agroecology. *Int. J. Agric. Sust.* **2009**, *7*, 3–18.

35. Wezel, A.; Bellon, S.; Doré, T.; Francis, C.; Vallod, D.; David, C. Agroecology as a science, a movement and a practice. A review. *Agron. Sustain. Dev.* **2009**, *29*, 503–515.

36. Altieri, M.A. Agroecology: The science of natural resource management for poor farmers in marginal environments.*Agric. Ecosys. Environ.* **2002**, *93*, 1–24.

37. Giampietro, M. *Multi-Scale Integrated Analysis of Agroecosystems: A Complex System Approach*; CRC press: Boca Raton, FL, USA, 2003.

38. Gomiero, T.; Giampietro, M.; Mayumi, K. Facing complexity on agro-ecosystems: A new approach to farming system analysis. *Int. J. Agric. Res. Gov. Ecol.* **2006**, *5*, 116–144.

39. Foley, J.A.; DeFries, R.; Asner, G.P.; Barford, C.; Bonan, G.; Carpenter, S.R.; Chapin, F.S.; Coe, M.T.; Daily, G.C.; Gibbs, H.K.; *et al.* Global consequences of land use. *Science* **2005**, *309*, 570–574.

40. Giampietro, M. Socioeconomic pressure, demographic pressure, environmental loading, and technological changes in agriculture. *Agric. Ecosys. Environ.* **1997**, *65*, 201–229.

41. Gomiero, T.; Giampietro, M.; Bukkens, S.M.; Paoletti, G.M. Biodiversity use and technical performance of freshwater fish culture in different socio-economic context: China and Italy. *Agric. Ecosys. Environ.* **1997**, *62*, 169–185.

42. Gomiero, T.; Giampietro, M. Multiple-scale integrated analysis of farming systems: The Thuong Lo commune (Vietnamese uplands) case study. *Pop. Environ.* **2001**, *22*, 315–352.

43. FAO. *Organic Agriculture, Environment and Food Security*; Environment and Natural Resources Service Sustainable Development Department, FAO: Rome, Italy, 2002. Available online: http://www.fao. org/ DOCREP/005/Y4137E/y4137e00.htm#TopOfPage (accessed on 12 January 2009).

44. UN (United Nations). Population Newsletter. Number 83. 2007. Availablenline: http://www.un.org/esa/population/publications/popnews/Newsltr_83.pdf (accessed on 2 July 2009).

45. Alexandratos, N.; Bruinsma, J. *World Agriculture towards 2030/2050*; The 2012 Revision, ESA Working Paper No. 12-03. FAO: Rome, Italy, 2012.Available online: http://www.fao. org/docrep/016/ ap106e/ap 106e.pdf (accessed on 24 June 2013).

46. Trostle, R. *Global Agricultural Supply and Demand: Factors Contributing to the Recent Increase in Food Commodity Prices*; United States Department of Agriculture: Washington, DC, USA, 2008. Available online: http://www. ers.usda.gov/PUBLICATIONS/ WRS0801/WRS0801. PDF (accessed on 20 May 2009).

47. UNEP (United Nations Environmental Programme). *The Environmental Food Crisis the Environment's Role in Averting Future Food Crises a UNEP Rapid Response Assessment*; UNEP, GRID-Arendal: Arendal, Norway, 2009. Available online:http:// www.grida.no/ publications/rr/ food-crisis/ (accessed on 12 May 2010).

48. FAO. The State of Food Insecurity in the Developing World. FAO: Rome, Italy, 2012. Available online: http://www.fao. org/docrep/ 016/ i3027e/ i3027e.pdf (accessed on 24 June 2010).

49. Vitousek, P.M.; Mooney, H.A.; Lubchenco, J.; Melillo, J.M. Human domination of earth's ecosystems. *Science* **1997**, *277*, 494–499.

50. Haberl, H.; Krausmann, F.; Erb, K.H.; Schulz, N.B.; Rojstaczer, S.; Sterling, S.M.; Moore, N. Human appropriation of net primary production. *Science* **2002**, *14*, 1968–1969.

51. Imhoff, M.L.; Bounoua, L.; Ricketts, T.; Loucks, C.; Harriss, R.; Lawrence, W.T. Global patterns in human consumption of net primary production. *Nature* **2004**, *429*, 870–873.

52. FAO. The State of the World's Land and Water Resources for Food and Agriculture (SOLAW)—Managing Systems at Risk. FAO: Rome, Italy, 2011. Available online: http://www.fao. org/docrep /017/i1688e/ i1688e.pdf (accessed on 9 August 2013).

53. FAO. *World Agriculture: Towards 2015/2030*; FAO: Rome, Italy, 2003. Available online: ftp://ftp.fao. org/docrep/fao /004/y3557e /y3557e. pdf (accessed on 20 January 2009).

54. Borlaug, N. Feeding a hungry world. *Science* **2007**, *318*, 359.

55. Godfray, H.C.J.; Crute, I.R.; Haddad, L.; Lawrence, D.; Muir, J.F.; Nisbett, N.; Pretty, J.; Robinson, S.; Toulmin, C.; Whiteley, R. The future of the global food system. *Phil. Trans. R. Soc. B* **2010**, *365*, 2769–2777.

56. Giampietro, M.; Mayumi, K. *The Biofuel Delusion: The Fallacy of Large Scale Agro-Biofuels Production*; Earthscan: London, UK, 2009.

57. Gomiero, T.; Paoletti, M.G.; Pimentel, D. Biofuels: Ethics and concern for the limits of human appropriation of ecosystem services. *J. Agric. Env. Ethics* **2010**, *23*, 403–434.

58. MacKay, D.J.C. Sustainable Energy—Without the Hot Air. 2009. Available online: http://www.withouthotair.com/download.html (accessed on 20 August 2009).

59. Giampietro, M.; Ulgiati, S.; Pimentel, D. Feasibility of large-scale biofuel production. *BioScience* **1997**, *47*, 587–600.

60. Rockstrom, J.; Lannerstad, M.; Falkenmark, M. Assessing the water challenge of a new green revolution in developing countries. *Proc. Natl. Acad. Sci. USA* **2007**, *104*, 6253–6260.

61. Smil, V. Water news: Bad, good and virtual. *Am. Sci.* **2008**, *96*, 339–407.

62. Krebs, J.R.; Wilson, J.D.; Bradbury, R.B.; Siriwardena, G.M. The second silent spring? *Nature* **1999**, *400*, 611–612.

63. Jarvis, D.I.; Hodgkin, T.; Sthapit, B.R.; Fadda, C.; Lopez-Noriega, I. An heuristic framework for identifying multiple ways of supporting the conservation and use of traditional crop varieties within the agricultural production system.*Crit. Rev. Plant. Sci.* **2011**, *30*, 125–176.

64. Zhu, Y.Y.; Chen, H.; Fan, J.; Wang, Y.; Li, Y.; Chen, J.; Fan, J.; Yang, S.; Hu, L.; Leung, H.; *et al.* Genetic diversity and disease control in rice. *Nature* **2000**, *406*, 718–722.

65. Green, R.E.; Cornell, S.J.; Scharlemann, J.P.W.; Balmford, A. Farming and the fate of wild nature. *Science* **2005**, *307*, 550–555.

66. Sachs, J.D.; Baillie, J.E.M.; Sutherland, W.J.; Armsworth, P.R.; Ash, N.; Beddington, J.; Blackburn, T.M.; Collen, B.; Gardiner, B.; Gaston, K.J.; *et al.* Biodiversity conservation and the Millennium Development Goals. *Science* **2009**, *325*, 1502–1503.

67. Fowler, C.; Hodgkin, T. Plant genetic resources for food and agriculture: Assessing global availability. *Annu. Rev. Environ. Resour.* **2005**, *29*, 143–179.

68. Cassman, K.G.; Dobermann, A.; Walters, D.T. Agroecosystems, nitrogen-use efficiency, and nitrogen management.*Ambio* **2002**, *31*, 132–140.

69. Erisman, J.W.; Sutton, M.A.; Galloway, J.; Klimont, Z.; Winiwarter, W. How a century of ammonia synthesis changed the world. *Nat. Geosci.* **2008**, *1*, 636–639.

70. Robertson, G.P.; Vitousek, P.M. Nitrogen in agriculture: Balancing the cost of an essential resource. *Annu. Rev. Environ. Resour.* **2009**, *34*, 97–125.

71. Birkhofer, K.; Schöning, I.; Alt, F.; Herold, N.; Klarner, B.; Maraun, M.; Marhan, S.; Oelmann, Y.; Wubet, T.; Yurkov, A.; *et al.* Gneral relationships between abiotic soil properties and soil biota across spatial scales and different Land-Use types. *PLoS One* **2012**, *7*, e43292.

72. Winston, M.K. *Nature Wars: People vs. Pests*; Harvard University Press: Cambridge, MA, USA, 1997.

73. Altieri, M.A.; Nicholls, C.I. *Biodiversity and Pest. Management in Agroecosystems*, 2nd ed.; CRC Press: Boca Raton, FL, USA, 2004.

74. *The Pesticide Detox*; Pretty, J., Ed.; Earthscan: London, UK, 2005.

75. Pimentel, D. *Technique for Reducing Pesticides Use. Economic and Environmental Benefits*; John Wiley & Sons: New York, NY, USA, 1997.

76. Richter, E.D. Acute Human Pesticide Poisonings. In *Encyclopedia of Pest. Management*; Pimentel, D., Ed.; Taylor & Francis: Boca Raton, FL, USA, 2002; pp. 3–6.

77. Colborn, T.; Dumanoski, D.; Meyers, J.P. *Our Stolen Future: How We Are Threatening Our Fertility, Intelligence and Survival. A Scientific Detective Story*; Penguin Books: New York, NY, USA, 1997.

78. Lyons, G. Effects of pollutants on the reproductive health of male vertebrate wildlife—Males under threat. *CHEM Trust (Chemicals, Health and Environment Monitoring)*, 2009. Available online: http://www.chemtrust. org.uk/documents/Male%20Wildlife%20 Under %20Threat%202008%20full%20report.pdf (accessed on 10 May 2010).

79. Stuart, T. *Waste: Uncovering the Global Food Scandal*; Penguin Books: London, UK, 2009.

80. Cuéllar, A.D.; Webber, M.E. Wasted Food, Wasted Energy: The embedded energy in Food waste in the United States.*Environ. Sci. Technol.* **2010**, *44*, 6464–6469.

81. FAO. Global Food Loses and Food Waste: Extent, Causes and Prevention. Study Conducted for the International Congress SAVE FOOD! at Interpack, Düsseldorf, Germany 2011. FAO: Rome, Italy, 2011. Available online: http://www.fao.org/docrep/ 014/mb060e/ mb060e00.pdf (accessed on 4 April 2013).

82. Pelletier, N.; Audsley, E.; Brodt, S.; Garnett, T.; Henriksson, P.; Kendall, A.; Kramer, K.J.; Murphy, D.; Nemecek, T.; Troell, M. Energy intensity of agriculture and food systems. *Annu. Rev. Environ. Resour.* **2011**, *36*, 223–246.

83. Smil, V. Eating meat: Evolution, patterns, and consequences. *Pop. Dev. Review* **2002**, *28*, 599–639.

84. Pimentel, D. Food for thought: A review of the role of energy in current and evolving agriculture. *Crit. Rev. Plant. Sci.***2011**, *30*, 35–44.

85. Pimentel, D.; Williamson, S.; Alexander, S.E.; Gonzalez-Pagan, O.; Kontak, C.; Mulkey, S.E. Reducing Energy Inputs in the US Food System. *Hum. Ecol.* **2008**, *36*, 459–471.

86. USDA (United State Development Agency). Food, Agriculture, Conservation, and Trade Act of 1990 (FACTA), Public Law 101-624, Title XVI, Subtitle A, Section 1603. Government Printing Office: Washington, DC, USA, 1990.

87. Jackson, W. Natural systems agriculture: A truly radical alternative. *Agric. Ecosys. Environ.* **2002**, *88*, 111–117.

88. Soule, J.; Piper, J. *Farming in Nature's Image: An Ecological Approach to Agriculture*; Island Press: Washington, DC, USA, 1991.

89. Odum, E.P. *Ecology and Our Endangered Life-Support. Systems*; Sinauer Publishers: Sunderland, MA, USA, 1993.

90. *Sustainable Agricultural Systems*; Edwards, C.A., Lal, R., Madden, P., Miller, R.H., House, G., Eds.; Soil and Water Conservation Society: Ankeny, LA, USA, 1990.

91. Pretty, J. *Agri.-Culture: Reconnecting People, Land and Nature*; Earthscan: London, UK, 2002.

92. Pretty, J. Agricultural sustainability: Concepts, principles and evidence. *Phil. Trans. B* **2008**, *363*, 447–465.

93. Pretty, J.N.; Noble, A.D.; Bossio, D.; Dixon, J.; Hine, R.E.; Penning derVires, F.V.T.; Morrison, J.I.L. Resource-conserving agriculture increases yields in developing countries. *Environ. Sci. Technol.* **2006**, *40*, 1114–1119.

94. Francis, C.; Lieblein, G.; Gliessman, S.; Breland, T.A.; Creamer, N.; Harwood, R.; Salomonsson, L.; Helenius, J.; Rickerl, D.; Salvador, R.; *et al*. Agroecology: The ecology of food systems. *J. Sust. Agric.* **2003**, *22*, 99–118.

95. Francis, C.A.; Porter, P. Ecology in sustainable agriculture practices and systems. *Crit. Rev. Plant. Sci.* **2011**, *30*, 64–73.

96. Gliessman, S.R. *Agroecology: Ecological Processes in Sustainable Agriculture*, 2nd ed.; Lewis Publisher: Boca Raton, FL, USA, 2007.

97. IPCC (Intergovernmental Panel on Climate Change). *Climate Change 2007: The Physical Science Basis. Contribution of Working Group I to the Fourth Assessment Report of the Intergovernmental Panel on Climate Change*; Cambridge University Press: Cambridge, UK, 2007. Available online: http://www.ipcc.ch/ (accessed on 20 May 2010).

98. U.S. EPA. *Inventory of U.S. Greenhouse Gas Emissions and Sinks: 1990–2008*; United States Environmental Protection Agency: Washington, DC,

USA, 2010. Available online: http://epa.gov/climatechange/emissions/downloads10/US-GHG-Inventory-2010_ExecutiveSummary.pdf (accessed on 14 May 2010).

99. Drinkwater, L.E.; Wagoner, P.; Sarrantonio, M. Legume-based cropping systems have reduced carbon and nitrogen losses. *Nature* **1998**, *396*, 262–265.

100. Schlesinger, W.H. Carbon and agriculture: Carbon sequestration in soils. *Science* **1999**, *284*, 2095.

101. Grandy, S.A.; Robertson, G.P. Land-use intensity effects on soil. *Ecosystems* **2007**, *10*, 58–73.

102. Smith, P.; Martino, D.; Cai, Z.; Gwary, D.; Janzen, H.; Kumar, P.; McCarl, B.; Ogle, S.; O'Mara, F.; Rice, C.; *et al*. Greenhouse gas mitigation in agriculture. *Phil. Trans. R. Soc. B* **2008**, *363*, 789–813.

103. Cox, T.S.; Glover, J.D.; van Tassel, D.L.; Cox, C.M.; Dehaan, L.R. Prospects for developing perennial grain crops.*Bioscience* **2005**, *59*, 649–659.

104. Glover, J.D.; Reganold, J.P.; Bell, L.W.; Borevitz, J.; Brummer, E.C.; Buckler, E.S.; Cox, C.M.; Cox, T.S.; Crews, T.E.; Culman, S.W.; *et al*. Increased food and ecosystem security via perennial grains. *Science* **2010**, *328*, 1638–1639.

105. Codex Alimentarius. Guidelines for the Production, Processing, Labelling and Marketing of Organically Produced Foods (GL 32–1999, Rev. 1–2001). 2004. Available online: http://www. codexalimentarius.net/web/standard_list.do?lang=en (accessed on 24 November 2007).

106. Courville, S. Organic Standards and Certification. In *Organic Agriculture. A global Perspective*; Kristiansen, P., Taji, A., Reganold, J., Eds.; CSIRO Publishing: Collingwood, Australia, 2006; pp. 201–220.

107. EC (European Commission). Council Regulation (EC) No 834/2007, of 28 June 2007 on organic production and labelling of organic products and repealing Regulation (EEC) No 2092. 2007. Available online: http://eurlex.europa.eu/LexUriServ/site/en/oj/2007/l_189/l_18920070720en00010023.pdf (accessed on 25 November 2009).

108. USDA (United States Department of Agriculture). *Agricultural Statistics*; U.S. Department of Agriculture: Washington, DC, USA, 2007.

109. IFOAM (International Movement of Organic Agriculture Movements), *The World of Organic Agriculture—Statistics and Emerging Trends 2008*; International Federation of Organic Agriculture Movements—IFOAM: Bonn, Germany, 2008.

110. IFOAM (International Movement of Organic Agriculture Movements). Definition of Organic Agriculture. IFOAM: Bon, Germany, 2010. Available online: http://www.ifoam.org (accessed on 10 March 2010).

111. Conford, P. *The Origins of the Organic Movement*; Floris Books: Glasgow, UK, 2001.

112. Lotter, D.W. Organic agriculture. *J. Sust. Agric.* **2003**, *21*, 59–128.

113. Heckman, J. A history of organic farming: Transitions from Sir Albert Howard's war in the soil to USDA National Organic Program. *Renew. Agr. Food Syst.* **2006**, *21*, 143–150.

114. *Organic Farming: An International History*, 2nd ed.; Lockeretz, W., Ed.; CABI: Wallingford, UK, 2007.

115. *Organic Agriculture. A Global Perspective*; Kristiansen, P., Taji, A., Reganold, J., Eds.; CSIRO Publishing: Collingwood, Australia, 2006.

116. Reganold, J.; Elliott, L.; Unger, Y. Long-term effects of organic and conventional farming on soil erosion. *Nature* **1987**,*330*, 370–372.

117. Lampkin, N. *Organic Farming*; Old Pond Publishing: Suffolk, UK, 2002.

118. Clark, M.S.; Horwath, W.R.; Shennan, C.; Scow, K.M. Changes in soil chemical properties resulting from organic and low-input farming practices. *Agron. J.* **1998**, *90*, 662–671.

119. Siegrist, S.; Staub, D.; Pfiffner, L.; Mäder, P. Does organic agriculture reduce soil erodibility? The results of a long-term field study on loess in Switzerland. *Agric. Ecosyst. Environ.* **1998**, *69*, 253–264.

120. Stockdale, E.A.; Lampkin, N.H.; Hovi, M.; Keatinge, R.; Lennartsson, E.K.M.; Macdonald, D.W.; Padel, S.; Tattersall, F.H.; Wolfe, M.S.; Watson, C.A. Agronomic and environmental implications for organic farming systems. *Adv. Agron***2001**, *70*, 261–327.

121. Mäder, P.; Flieβbach, A.; Dubois, D.; Gunst, L.; Fried, P.; Niggli, U. Soil fertility and biodiversity in organic farming.*Science* **2002**, *296*, 1694–1697.

122. Lotter, D.W.; Seidel, R.; Liebhart, W. The performance of organic and conventional cropping systems in an extreme climate year. *Am. J. Altern. Agr.* **2003**, *18*, 146–154.

123. Bengtsson, J.; Ahnstrom, J.; Weibull, A.C. The effects of organic agriculture on biodiversity and abundance: A meta-analysis. *J. Appl. Ecol.* **2005**, *42*, 261–269.

124. Pimentel, D.; Hepperly, P.; Hanson, J.; Douds, D.; Seidel, R. Environmental, energetic, and economic comparisons of organic and conventional farming systems. *Bioscience* **2005**, *55*, 573–582.

125. Birkhofer, K.; Bezemer, T.M.; Bloem, J.; Bonkowski, M.; Christensen, S.; Dubois, D.; Ekelund, F.; Flieβbach, A.; Gunst, L.; Hedlund, K.; *et al*. Long-term organic farming fosters below and aboveground biota: Implications for soil quality, biological control and productivity. *Soil Biol. Biochem.* **2008**, *40*, 2297–2308.

126. Briar, S.S.; Grewal, P.S.; Somasekhar, N.; Stinner, D.; Miller, S.A. Soil nematode community, organic matter, microbial biomass and nitrogen dynamics in field plots transitioning from conventional to organic management. *Appl. Soil. Ecol.***2007**, *37*, 256–266.

127. Flieβbach, A.; Oberholzer, H.R.; Gunst, L.; Mäder, P. Soil organic matter and biological soil quality indicators after 21 years of organic and conventional farming. *Agric. Ecosyst. Environ.* **2007**, *118*, 273–284.

128. Liu, B.; Tu, C.; Hu, S.; Gumpertz, M.; Ristaino, J.B. Effect of organic, sustainable, and conventional management strategies in grower fields on soil physical, chemical, and biological factors and the incidence of Southern blight. *Appl. Soil Ecol.* **2007**, *37*, 202–214.

129. Teasdale, J.R.; Coffman, C.B.; Mangum, R.W. Potential long-term benefits of no-tillage and organic cropping systems for grain production and soil improvement. *Agron. J.* **2007**, *99*, 1297–1305.

130. Gattinger, A.; Muller, A.; Haeni, M.; Skinner, C.; Fliessbach, A.; Buchmann, N.; Mäder, P.; Stolze, M.; Smith, P.; El-Hage Scialabbad, N.; *et al.* Enhanced top soil carbon stocks under organic farming. *Proc. Natl. Acad. Sci. USA* **2012**, *109*, 18226–18231.

131. Chivenge, P.; Vanlauwe, B.; Six, J. Does the combined application of organic and mineralnutrient sources influence maize productivity? A meta-analysis. *Plant Soil* **2011**, *342*, 1–30.

132. Kramer, S.B.; Reganold, J.P.; Glover, J.D.; Bohannan, B.J.M.; Mooney, H.A. Reduced nitrate leaching and enhanced denitrifier activity and efficiency in organically fertilized soils. *Proc. Natl. Acad. Sci. USA* **2006**, *103*, 4522–4527.

133. Küstermann, B.; Christen, O.; Hülsbergen, K.J. Modelling nitrogen cycles of farming systems as basis of site- and farm-specific nitrogen management. *Agric. Ecosys. Environ.* **2010**, *135*, 70–80.

134. Sieling, K.; Kage, H. N balance as an indicator of N leaching in an oilseed rape-winter wheat winter barley rotation.*Agric. Ecosys. Environ.* **2006**, *115*, 261–269.

135. Kirchmann, H.; Bergström, L.; Kätterer, T.; Mattsson, L.; Gesslein, S. Comparison of long-term organic and conventional crop–livestock systems on a previously nutrient-depleted soil in Sweden. *Agron. J.* **2007**, *99*, 960–972.

136. Wu, J.Y.; Sardo, V. Sustainable *versus* Organic Agriculture. In *Sociology, Organic Farming, Climate Change and Soil Science*; Lichtfouse, E., Ed.; Springler: Dordrecht, Germany, 2010; pp. 41–76.

137. Stanhill, G. The comparative productivity of organic agriculture. *Agric. Ecosyst. Environ.* **1990**, *30*, 1–26.

138. Smolik, J.D.; Dobbs, T.L.; Rickerl, D.H. The relative sustainability of alternative, conventional and reduced-till farming system. *Am. J. Altern. Agr.* **1995**, *10*, 25–35.

139. Teasdale, J.R.; Rosecrance, R.C.; Coffman, C.B.; Starr, J.L.; Paltineanu, I.C.; Lu, Y.C.; Watkins, B.K. Performance of reduced-tillage cropping systems for susta nable grain production in Maryland. *Am. J. Altern. Agr.* **2000**, *15*, 79–87.

140. Fuller, R.J.; Norton, L.R.; Feber, R.E.; Johnson, P.J.; Chamberlain, D.E.; Joys, A.C.; Mathews, F.; Stuart, R.C.; Townsend, M.C.; Manley, W.J.; *et al.* Benefits of organic farming to biodiversity vary among taxa. *Biol. Lett.* **2005**, *1*, 431–434.

141. Hole, D.G.; Perkins, A.J.; Wilson, J.D.; Alexander, I.H.; Grice, P.V.; Evans, A.D. Does organic farming benefit biodiversity? *Biol. Conserv.* **2005**, *122*, 113–130.

142. Phelan, P.L.; Mason, J.F.; Stinner, B.R. Soil-fertility management and host preference by European corn borer, *Ostrinia. nubilalis* (Hübner), on *Zea. mays* L.: A comparison of organic and conventional chemical farming. *Agric. Ecosys. Environ.* **1995**, *56*, 1–8.

143. Phelan, P.L.; Norris, K.H.; Mason, J.F. Soil-management history and host preference by *Ostrinia nuhilalis*: Evidence for plant mineral balance mediating insect-plant interactions. *Environ. Entomol.* **1996**, *25*, 1329–1336.

144. Alyokhin, A.; Porter, G.; Groden, E.; Drummond, F. Colorado potato beetle response to soil amendments: A case in support of the mineral balance hypothesis? *Agric. Ecosys. Environ.* **2005**, *109*, 234–244.

145. Hsu, Y.T.; Shen, T.C.; Hwang, S.Y. Soil fertility management and pest responses: A compar son of organic and synthetic fertilization. *J. Econ. Entomol.* **2009**, *102*, 160–169.

146. Staley, J.T.; Stewart-Jones, A.; Pope, T.W.; Wright, D.J.; Leather, R.S.; Hadley, P.; Rossiter, J.T.; van Emden, H.F.; Poppy, G.M. Varying responses of insect herbivores to altered plant chemistry under organic and conventional treatments. *Proc. R. Soc. B* **2010**, *277*, 779–786.

147. Butler, J.; Garratt, M.P.D.; Leather, S.R. Fertilisers and insect herbivores: A meta-analysis. *Ann. Appl. Biol.* **2012**, *161*, 223–233.

148. Garratt, M.P.D.; Wright, D.J.; Lather, S.R. The effects of farming system and fertilisers on pests and natural enemies: A synthesis of current research. *Agric. Ecosyst. Environ.* **2011**, *141*, 261–270.

149. Reganold, J.P. The fruits of organic farming. *Nature* **2012**, *485*, 176.

150. Fierer, N.; Grandy, A.S.; Six, J.K.; Paul, E.A. Searching for unifying principles in soil ecology. *Soil Biol. Biochem.* **2009**,*41*, 2249–2256.

151. Young, A. *Land Resources: Now and for the Future*; Cambridge University Press: Cambridge, UK, 1998.

152. Perfecto, I.; Vandermeer, J.; Wright, A. *Nature's Matrix. Linking Agriculture, Conservation and Food Sovereignty*; Earthscan: London, UK, 2009.

153. Vandermeer, J.; Perfecto, I.; Philpott, S. Ecological complexity and pest control in organic coffee production: Uncovering an autonomous ecosystem service. *BioScience* **2010**, *60*, 527–537.

154. Tscharntke, T.; Clough, Y.; Wanger, T.C.; Jackson, L.; Motzke, I.; Perfecto, I.; Vandermeer, J.; Whitbread, A. Global food security, biodiversity conservation and the future of agricultural intensification. *Biol. Conserv.* **2012**, *151*, 53–59.

155. De Ponti, T.; Rijk, B.; van Ittersum, M.K. The crop yield gap between organic and conventional agriculture. *Agric. Syst.***2012**, *108*, 1–9.

156. Seufert, V.; Ramankutty, N.; Foley, J.A. Comparing the yields of organic and conventional agriculture. *Nature* **2012**,*485*, 229–232.

157. Beets, W.C. *Raising and Sustaining Productivity of Smallholder Farming System in the Tropics*; AgBe Publishing: Alkmaar, The Netherlands, 1990.

158. Leopold, A. *A Sand County Almanac*; Oxford University Press: New York, NY, USA, 1949.

# Chapter 9

# Scenario Studies on Effects of Soil Infiltration Rates, Land Slope, and Furrow Irrigation Characteristics on Furrow Irrigation-Induced Erosion

**Jibrin M. Dibal,[1] A. A. Ramalan,[2] O. J. Mudiare,[2] and H. E. Igbadun[2]**

[1]Department of Agricultural and Environmental Resources Engineering, Faculty of Engineering, University of Maiduguri, PMB 1069, Maiduguri, Borno State, Nigeria

[2]Department of Agricultural Engineering, Faculty of Engineering, Ahmadu Bello University, PMB 1044, Zaria, Nigeria

## ABSTRACT

Furrow irrigation proceeds under several soil-water-furrow hydraulics interaction dynamics. The soil erosion consequences from such interactions in furrow irrigation in Samaru had remained uncertain. A furrow irrigation-induced erosion (FIIE) model was used to simulate the potential severity of soil erosion in irrigated furrows due to interactive effects of infiltration rates, land slope, and some furrow irrigation characteristics under different scenarios. The furrow irrigation characteristics considered were furrow lengths, widths, and stream sizes. The model itself was developed using the dimensional analysis approach. The scenarios studied were the interactive effects of furrow lengths, furrow widths, and slopes steepness; infiltration rates and furrow lengths; and stream sizes, furrow lengths, and slopes steepness on potential furrow irrigation-induced erosion, respectively. The severity of FIIE was found to relate somewhat linearly with slope and stream size, and inversely with furrow lengths and furrow width. The worst soil erosion (378.05 t/ha/yr) was found as a result of the interactive effects of 0.65 m furrow width, 50 m furrow length, and 0.25% slope steepness; and the least soil erosion (0.013 t/ha/yr) was induced by the combined effects of 0.5 l/s, 200 m furrow length, and 0.05% slope steepness. Evidently considering longer furrows in furrow irrigation designs would be a better alternative of averting excessive FIIE.

## 1. INTRODUCTION

The use of mathematical models and computer simulations in engineering, hydrology, and various fields in the process of decision making is progressively gaining more acceptances [1]. Models are useful tools for representing the detailed and complex "real world" with a more simple and understandable structure and can be used in demonstrating the relationships and interactions amongst various factors [2]. Furthermore, models allow the decision-maker to combine information from various sources and, in some cases, to extrapolate findings beyond the trial period [1]. They can also be employed in the identification of promising technologies and reduce the time and expense of field experiments by focusing resources on these most promising technologies [3].

Furrow irrigation is one of the most conventionally practiced irrigation methods in the area. About 60% of irrigated lands in Nigeria are furrow-irrigated [4]. Furrow irrigation is especially recommended for growing row crops on medium- to heavy-textured soils due to its simplicity and low capital costs [5]. Compared to traditional strip plot and other surface irrigation methods, furrow irrigation was found to have reduced water requirements by up to 28–43 percent [4]. However, one of the drawbacks of furrow irrigation is the soil erosion that it occasions. By this process, nutrient-rich top soil is removed, transported, and deposited somewhere, thereby constituting a nuisance. Soil erosion impacts negatively on both the environment and crop productivity and depresses the economic and social status of farmers [6]. Sojka et al. [7] reported 75% of Idaho furrow-irrigated fields lost their entire " A" horizon in the upper reaches together with a 2- to 4-fold increase in "topsoil" at the lower ends, reducing productivity by 25% over preerosion values and reducing yields by 20–50% in areas where top soil has been lost. Furrow irrigation-induced erosion (FIIE) has also become one of the major factors limiting factors to sustainable irrigated crop production in Samaru.

Erosion problems are complex owing to a combination of socioeconomic factors and common agricultural practices. A wide range of locally and internationally funded research projects from the government, nongovernmental organizations (NGOs), and the academia have addressed rainfall erosion problems in Samaru, for example, the works of Adewumi [8], Adewumi et al. [9], and Odunze [10]. But the predicament of FIIE has not received adequate attention. Unfortunately, the technologies and methods developed to understand, predict, and/or mitigate soil erosion that focused on rainfall-induced erosion cannot be effective on irrigation-induced erosion [11]. There is therefore a growing need for models to be applied in predicting, control, and understanding of FIIE. Leib et al. [12] pointed out that the minimum information needed to evaluate soil erosion in furrow includes the amount of soil lost during the irrigation along with some field and irrigation characteristics such as field slope, field length, soil texture, furrow inflow rate, and irrigation duration. Other useful information is furrow spacing, hydraulic radius, and soil-water continuum.

The discharge measurements which are the basis of FIIE assessment are commonly used to calculate the average soil loss per unit field area (Mg/ha) (Bjorneberg and Strelkoff, 2001). However, from a practical viewpoint, erosion rates on a field vary widely with many soil and furrow hydraulic factors. For example, erosion and sediment transport capacity increase with the shear or velocity of the flow, which in turn increase with the flow rate and furrow slope. Further, 50 to 80% of a typical furrow inflow infiltrates before it reaches the furrow end, resulting in a corresponding flow rate decrease along the furrows [13]. Understanding the interactions of soil and hydraulic factors affecting FIIE is indispensable for satisfactory decision making and management. It is obvious therefore that the future context of FIIE that an individual, an organization, or a development policy will have to respond to is both complex and uncertain. The ability to visualize and simulate support on problems related to land-soil-furrow hydraulics interactions through negotiating alternative practical possible combination of possible events or scenarios and proffer solution to them simultaneously is a vital tool in research toward averting FIIE.

Nicetic et al. [14] showed that scenarios studies can be used to generate knowledge about the present and the future and to identify the limits of that knowledge. It is also used as a public communication tool to draw the attention of the public to specific issues and for examining the potential effectiveness of a decision in solving a particular problem or understanding the potential dangers associated with interaction of working variables. Ajayi and Horta [15] used scenario analysis to investigate the effect of spatially varied soil hydraulic properties on runoff. This study aims at exploring the potential interactive effects of soil infiltration rates, slope steepness, and furrow irrigation characteristics on furrow irrigation-induced erosion in a semiarid region using a FIIE model. A brief of the model development is also presented.

# 2. MATERIALS AND METHODS

## 2.1. Study Area

Field experiments were conducted during the 2009/2010 and 2010/2011 irrigation seasons at the Irrigation Research Field of the Institute for Agricultural Research (IAR) farm, Samaru, Zaria, along the Zaria-Sokoto road (11°1′N, 7°38′E, on the altitude of 686 m above mean sea level). Samaru is situated within the Northern Guinea savanna zone of Nigeria.

## 2.2. Experimental Treatments

The experimental factors studied were stream size *Q*, furrow length *L*, and furrow widths *W* at 3, 2, and 2 levels, respectively. The stream sizes were 2.5, 1.5, and 0.5 l/s; furrow lengths were 90 and 45 m; and furrow widths were 0.75 and 0.9 m. The combination of the stream sizes, furrow length, and furrow widths resulted in twelve (12) different treatments that were imposed on the field as follows:

$$\begin{aligned} &T_1 = Q_1L_1W_1, && T_2 = Q_1L_1W_2, \\ &T_3 = Q_1L_2W_1, && T_4 = Q_1L_2W_2, \\ &T_5 = Q_2L_1W_1, && T_6 = Q_2L_1W_2, \\ &T_7 = Q_2L_2W_1, && T_8 = Q_2L_2W_2, \\ &T_9 = Q_3L_1W_1, && T_{10} = Q_3L_1W_2, \\ &T_{11} = Q_3L_2W_1, && T_{12} = Q_3L_2W_2, \end{aligned} \tag{1}$$

where $Q_1$, $Q_2$, and $Q_3$ are stream sizes 2.5, 1.5, and 0.5 l/s, respectively; $L_1$ and $L_2$ are furrow lengths 90 and 45 m, respectively; $W_1$ and $W_2$ are furrow widths 0.75 and 0.9 m. The layout of the experiment was a randomized complete block laid in a split plot design with four replications. This layout was observed in both 2009/2010 (Trial 1) and 2010/2011 (Trial 2) seasons. Main plots consisted of stream sizes while furrow lengths and furrow widths were focused in the subplots. In both of the study seasons, each replication comprised three plots, and each experimental plot had three and two ridges in Trials 1 and 2, respectively.

## 2.3. Field Measurements

### 2.3.1. Preirrigation Data

The basic infiltration rate of the soil was determined using the inflow-outflow method. Slope of the field was determined in each of the two trials using the dumpy level surveying instrument. The hydraulic shear stress $\tau$ was obtained following the guidance of Schwab et al. [16]. The soil erodibility, $K$, was determined by adopting the universal soil loss equation (USLE) "$K$" equation as it is in Wall et al. [17]. The time of flow of water was measured directly from the field using a stopwatch.

### 2.3.2. Irrigation/Erosion Related Data

Prior to commencement of irrigations, water-sediment collection stations were established 5 m before the end of each furrow. Flow of water in the furrows was measured using a cutthroat flume installed 5 m from entry at the upstream and at the tail end of the furrows for the measurement of water inflow and outflow of the furrows. Water flowing out of the furrows was measured as runoff. One liter of water-sediment samples was collected during irrigation events at each of the established measurement points and was taken to laboratory and filtered into preweighed metal containers. The resultant residues were oven-dried at 105°C over 24-hour period and reweighed. Sediment concentrations (g/L) were calculated from the dried residues and the runoff volumes for each of the treatments and were used to calculate soil erosion per furrow. Runoff volume was calculated as the product of the runoff discharge (l/s) (from the downstream flumes) and duration of runoff discharge. Soil erosion at the end of the furrows was calculated as the product of the sediment concentrations and runoff volumes divided by the wetted area. The wetted area was calculated as the product of the top width of flow of water and the length of the furrow [6, 18–20].

## 2.4. Model Development

The furrow irrigation-induced erosion (FIIE) model was developed using dimensional analysis approach following the concept of Buckingham π-theorem considering the variables in Table 1.

**Table 1.** Variables considered in the model development.

| Variables | Symbols | Units | Dimensional symbols (M, L, T) |
|---|---|---|---|
| Soil erosion | $E$ | $Kg/m^2/yr$ | $ML^{-2}T^{-1}$ |
| Stream size | $Q$ | $m^3/s$ | $L^{-3}T^{-1}$ |
| Slope | $S$ | % | [] |
| Furrow lengths | $X$ | m | L |
| Top widths of flow | $W$ | m | L |
| Hydraulic radius | $R$ | m | L |
| Infiltration rate | $I$ | mm/hr | $LT^{-1}$ |
| Time of flow | $T$ | sec | T |
| Flow velocity | $V$ | m/s | $LT^{-1}$ |
| Manning's roughness coefficient | $n$ | $sec/m^{1/3}$ | $TL^{-1/3}$ |
| Acceleration due to gravity | $g$ | $m/s^2$ | $LT^{-2}$ |
| Density of water | $\rho_w$, | $kg/m^3$ | $ML^{-3}$ |
| Soil particle density | $\rho_s$, | $kg/m^3$ | $ML^{-3}$ |

However, the principles of the Buckingham π-theorem require that the variables to be used in the dimensional analysis be theoretically independent. Some of the variables in (1) were therefore eliminated, and the resultant is the decision variables used in the model development as presented in Table 2.

**Table 2.** Decision variables and their corresponding dimensions.

| Variables | Symbols | Units | Dimensional symbols (M, L, T) |
|---|---|---|---|
| Soil erosion | $E$ | kg/m$^2$/yr | $ML^{-2}T^{-1}$ |
| Stream size | $Q$ | m$^3$/s | $L^3T^{-1}$ |
| Furrow length | $X$ | m | L |
| Top width of flow | $W$ | m | L |
| Soil infiltration rate | $I$ | mm/hr | $LT^{-1}$ |
| Time of flow of water | $T$ | s | T |
| Soil erodibility factor | $K$ | Kg·hr/Nm$^2$ | $L^{-3}T^3$ |
| Hydraulic shear stress | $\tau$ | N/m$^2$ | $ML^{-1}T^{-2}$ |

A two-year field experiment was embarked upon to collect data for the completion of the model development. Five π-terms, namely, $\pi_1$ to $\pi_5$, were developed using the dimensional analysis and later combined to three. Three graphs were plotted, namely, $\pi_1$ against $\pi_2$ keeping $\pi_3$ constant, $\pi_1$ against $\pi_3$ keeping $\pi_2$ constant, and $\pi_1$ against $\pi_2$ keeping $\pi_3$ constant with supplementary data. Component equations were derived from the regression equations of the three graphs and they were further used to develop the model as presented in

$$E = 0.1110\left(\frac{\tau}{I}\right) + 0.4117\left(\frac{TQ^{1/2}\tau}{XWIK^{1/6}}\right) - 0.0039\left(\frac{\tau QK^{1/3}}{XWI}\right). \quad (2)$$

The validation of the model showed that the model's prediction has a 72.68% degree of accuracy (Figure 1). The efficiency criteria (ECs) [21, 22] used to study the error margin between the simulated and observed FIIE values showed the model has coefficient of variation (CV) of 0.412, Nash-Sutcliffe efficiency (NSE) of 0.7, and index of agreement (*D*) of 0.941. The model was found to be most sensitive to stream size and least sensitive to time of flow of water in the furrows.

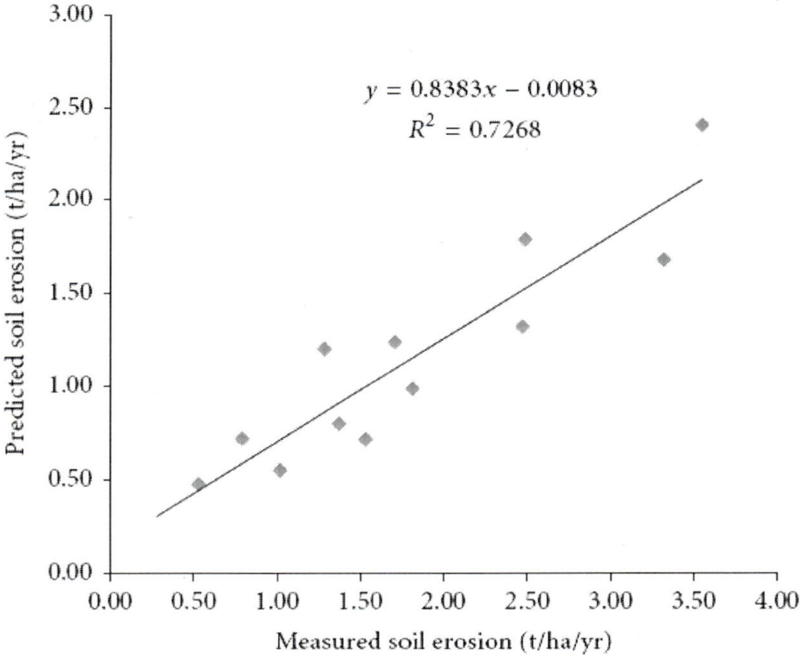


**Figure 1.** Comparison of predicted with measured soil erosion.

## 2.5. Scenario Studies

Soil erosion in irrigated furrow is practically a function of many variables, among which only stream sizes, furrow length, and width were studied, and it is often not possible to monitor the influence of every land-use practice under all weather and soil conditions. Scientific planning for soil conservation and water management requires knowledge of the relations among those parameters that cause the soil erosion [23]. These parameters can be found by a careful analysis of the process or by controlled experimentation. Soil erosion predictions were therefore performed using the model and were used to rank alternative practices (scenarios) with regard to their likely cause and/or severity of soil erosion. In each of the scenarios, the selected values were inserted into the model and the resultant FIIE was computed using the model.

Scenario 1: Effects of Furrow Lengths, Furrow Widths, and Slopes Steepness on Potential Furrow Irrigation-Induced Erosion. According to Savva and Frenken [24], the width of the furrows varies from 45 to 65 cm, and the furrow spacing varies between 0.75 and 1.0 m, depending on soil type, crops, and stream size to be applied to the furrow. While coarse soils require closely spaced furrows in order to achieve lateral water flow in the root zone, clayey soils are comparatively spaced wider. These dynamics lead to variations in top width of flow and depth flow of water in the furrows during irrigation events. Conversely, typical furrow lengths vary from about 60 m on coarse textured soils to 500 m on fine textured soils, depending on the land slope, stream size, and irrigation depth. In areas where there is a considerable risk of erosion, the maximum slope should be limited to 0.3% [25]. Low irrigation efficiencies are usually associated with poor land leveling, erroneous stream size selection, and changes in soil type along the irrigated area both vertically and horizontally. However, the potential soil erosion corresponding to any combination of slope, top width of flow, and furrow length is still vague. But a large number of irrigations are going on with these types of combinations, implying a large potential of soil erosion threat. In this scenario study, to have a close representation of variation of top widths of flow due to changes of furrow widths, four furrow widths (0.65, 0.75, 0.85, and 0.95 m) were selected. Four furrow lengths (50, 100, 150, and 200 m) also selected have a near representation of various furrow lengths used by farmers in Nigeria. Slope steepness of 0.3, 0.5, 1, 1.5, 2, and 2.5% was considered to cover the acceptable slope steepness $S$ in furrow irrigation.

Scenario 2: Effect of Combined Infiltration Rates and Furrow Lengths on Potential Furrow Irrigation-Induced Erosion. A number of soil factors affect farm irrigation system selection. Among other factors, soil texture plays a vital role through its effect on the available soil moisture and, most importantly, the infiltration rate of the soil. Many shallow soils, often with a depth of no more than 30 cm, are commonly developed for irrigation in many regions. At times, such soils may be of light texture obliging very frequent irrigation. The infiltration rate affects the length of run, size of furrows, and the application rates. Within the same soil textural group, infiltration rate can vary considerably, for instance, 20 to 30 and 5 to 10 mm/hr for sandy loam and clay loam soils, respectively [24]. Irrigation systems should be designed and managed in a way that the water application rate does not exceed the infiltrability of the soil. This scenario looks at potential erosion corresponding to furrow irrigation in various lengths of runs under varied infiltration rates.

Scenario 3: Effects of Combination of Stream Sizes, Furrow Lengths, and Field Slopes on Potential Furrow Irrigation-Induced Erosion. In irrigation practice, the use of very small stream size, prolongs infiltration opportunity time (I.O.T.) and causes deep percolation at the upstream end of the furrow if the bottom end is also to receive enough water. On the other hand, the use of very large stream size speeds up the waterfront and reduces I.O.T.; this leads to high runoff losses and erosion hazards in the furrows. Apparently, the selection of appropriate combinations of stream sizes and furrow length is necessary. The I.O.T. can be regulated in irrigated furrow to avert over- or underirrigation and its corresponding negative implications. Practically, this has not been of any concern to farmers despite its necessity in achieving good irrigation efficiency and averting erosion hazards. Water application can be regulated by increasing or decreasing the I.O.T. One can, within limits, regulate the depth of water applied. One method of achieving this is by decreasing the I.O.T. by steepening the slope $S$ and shortening the furrow length $l$ or by increasing the stream flow. In this method, one factor is varied while two other factors remain unchanged. But what is the magnitude of erosion problem associated with the combination of any factors mentioned? This scenario addresses it. In this analysis, three stream sizes (0.5, 1.5, and 2.5 l/s), four furrow lengths (50, 100, 150, and 200 m), and four slope steepness values (0.05, 0.3, 0.25, and 0.2%) were selected as representatives of what is commonly found in farmers' farm. For every value of $Q$, a corresponding value of and was measured from the field during the experiment and the values of were computed therefrom. The model was then used to generate the potential soil erosion values for every combination of $Q$, $l$ , and . Table 3 gives some of the hydraulic properties measured.

**Table 3.** Potential furrow irrigation-induced erosion (t/ha/yr) at combined furrow lengths, furrow widths, and slopes steepness.

| Furrow widths (m) | Furrow lengths (m) | Slopes (%) | | | | |
|---|---|---|---|---|---|---|
| | | 0.05 | 0.1 | 0.15 | 0.2 | 0.25 |
| | | Furrow irrigation-induced erosion (t/ha/yr) | | | | |
| 0.65 | 50 | 75.6111 | 151.3 | 226.83 | 302.44 | 378.05 |
| | 100 | 38.8 | 77.6 | 116.40 | 155.20 | 194.00 |
| | 150 | 26.52 | 50.06 | 79.59 | 106.12 | 132.65 |
| | 200 | 20.4 | 40.8 | 61.19 | 81.58 | 101.98 |
| 0.75 | 50 | 65.79 | 131.59 | 197.38 | 263.18 | 328.97 |
| | 100 | 33.89 | 67.78 | 101.68 | 135.57 | 169.46 |
| | 150 | 23.26 | 46.52 | 69.78 | 93.04 | 116.30 |
| | 200 | 17.94 | 35.88 | 53.82 | 71.77 | 89.71 |
| 0.85 | 50 | 58.28 | 116.56 | 174.85 | 233.13 | 291.42 |
| | 100 | 30.13 | 60.27 | 90.41 | 120.54 | 150.68 |
| | 150 | 20.75 | 41.50 | 62.26 | 83.01 | 103.77 |
| | 200 | 16.06 | 32.12 | 48.18 | 64.25 | 80.31 |
| 0.95 | 50 | 52.34 | 104.69 | 157.04 | 209.39 | 261.74 |
| | 100 | 27.16 | 54.33 | 81.50 | 108.67 | 135.84 |
| | 150 | 18.77 | 37.55 | 56.32 | 75.10 | 93.87 |
| | 200 | 14.57 | 29.15 | 43.73 | 58.31 | 72.89 |

## 3. RESULTS AND DISCUSSION

### 3.1. Combined Effects of Furrow Lengths, Furrow Widths, and Slopes Steepness on Potential Furrow Irrigation-Induced Erosion

Table 3 contains the potential furrow irrigation-induced soil erosion resulting from different choices of furrow widths, lengths, and slopes. The table shows that, at any particular furrow width and length, there could be a very severe danger of soil erosion if there is any change in slope along the length of run. The highest soil erosion (378.05 t/ha/yr) was found as a result of the combined effects of irrigating in furrows that were 50 m long and 0.65 m wide on 0.25% steep fields. This prediction stresses the need of farm leveling to ensure uniform slope prior to commencement of irrigation. Practically, many farmers do not consider farm leveling as an important issue in irrigation, possibly due to illiteracy or fear of cost. The trend, however, cautioned against the practice of irrigation in short furrows irrespective of the slope in use.

## 3.2. Effect of Combined Infiltration Rates and Furrow Lengths on Potential Furrow Irrigation-Induced Erosion

The predicted FIIE as consequent combinations of furrow lengths and various infiltration rates is presented in Table 4. The table shows a diminishing trend of soil erosion with increase in furrow lengths at any particular infiltration rate. For example, at 20 mm/hr, soil erosion would dwindle by 59% between 50 and 200 m furrow lengths. This supports the possibility of irrigation in long furrows with little fear of severe soil erosion and counsels against irrigation in short furrows.

**Table 4.** Potential furrow irrigation-induced erosion (t/ha/yr) at combined infiltration rates and furrow lengths.

| Furrow lengths (m) | Basic infiltration rates (mm/hr) | | | | |
|---|---|---|---|---|---|
| | 20 | 22 | 24 | 28 | 30 |
| 50 | 1.867 | 1.697 | 1.556 | 1.333 | 1.244 |
| 100 | 1.134 | 1.031 | 0.945 | 0.810 | 0.756 |
| 150 | 0.890 | 0.809 | 0.742 | 0.636 | 0.594 |
| 200 | 0.768 | 0.698 | 0.640 | 0.549 | 0.512 |

Similarly, the prediction shows another decreasing trend of soil erosion at a particular furrow length with increase in infiltration rate. For example, soil erosion dropped by about 33% in any particular furrow length between infiltration rates of 20 to 30 mm/hr. Often many irrigated fields, especially large farms, could have more than one type of soils or some slight spatial variation in the composition of the soil, hence varying in infiltration rates. The table supplies the irrigator with estimate possible of soil erosion that could be incurred and could be useful for planning irrigation.

## 3.3. Effects of Combination of Stream Sizes, Furrow Lengths, and Field Slopes on Potential Furrow Irrigation-Induced Erosion

Table 5 shows predicted furrow irrigation-induced erosion corresponding to practical combination of stream sizes, furrow lengths, and field slopes. The information that soil erosion increases with slope as it is in the table is not new; the table, however, provides the potential magnitude of soil erosion for any choice.

The prediction also called for prudence to be exercised when selecting large stream sizes to irrigate short furrows on steep slopes. When it is necessary to irrigate steep furrows, smaller stream size in a longer length of run is a better and safer option as pointed out in the table. The information in Tables 3 through 5 would serve as guide for the determination of optimum combinations of furrow lengths, widths, and stream sizes under varied slope and infiltration rates.

# 4. CONCLUSION

Models are employed in the systematic analysis of complex causal relationships to simulate the reactions and interactions of different variables to be able to identify inexplicit causal relationships of variables and their consequences. The FIIE model was herein used to explore the soil erosion consequences of the combinations of field slope, furrow lengths, top width of flow, infiltration rates, and stream sizes at various scales during furrow irrigation. The result stressed the danger of irrigating on steep slopped lands, cautioned against irrigation in short furrows, and called for prudent selection of flow stream sizes to circumvent severe soil erosion during furrow irrigation. Conducting such visualizations on a routine basis will help in providing a roadmap towards realization of sustainable irrigated agriculture.

**Table 5.** Predicted furrow irrigation-induced erosion (t/ha/yr) corresponding to combination of stream sizes, furrow lengths, and field slopes.

| Stream size (l/s) | Furrow lengths (m) | Slope (%) | | | |
|---|---|---|---|---|---|
| | | 0.05 | 0.10 | 0.15 | 0.20 |
| 0.5 | 50 | 0.024 | 0.047 | 0.071 | 0.095 |
| | 100 | 0.016 | 0.033 | 0.049 | 0.066 |
| | 150 | 0.014 | 0.027 | 0.041 | 0.054 |
| | 200 | 0.013 | 0.026 | 0.039 | 0.052 |
| 1.5 | 50 | 0.030 | 0.061 | 0.091 | 0.121 |
| | 100 | 0.019 | 0.039 | 0.058 | 0.078 |
| | 150 | 0.015 | 0.030 | 0.045 | 0.060 |
| | 200 | 0.014 | 0.028 | 0.043 | 0.057 |
| 2.5 | 50 | 0.043 | 0.086 | 0.128 | 0.171 |
| | 100 | 0.027 | 0.053 | 0.080 | 0.106 |
| | 150 | 0.020 | 0.040 | 0.060 | 0.080 |
| | 200 | 0.019 | 0.038 | 0.056 | 0.075 |

## REFERENCES

1. R. B. Foltz, W. J. Elliot, and N. S. Wagenbrenner, "Soil erosion model predictions using parent material/soil texture-based parameters compared to using site-specific parameters," Transactions of the ASABE, vol. 54, no. 4, pp. 1347–1356, 2011.

2. D. L. Bjorneberg, D. C. Kincaid, R. D. Lentz, R. E. Sojka, and T. J. Trout, "Unique aspects of modeling irrigation-induced soil erosion,"

International Journal of Seciment Research, vol. 15, no. 2, pp. 245–252, 2000.

3. C. J. Skonard and D. L. Martin, "A physically based two-dimensional infiltration model for furrow irrigation biological systems engineering conference presentations and white papers," ASAE Paper022063, 2002.

4. T. O. Dauda, O. E. Asiribo, S. O. Akinbode, J. O. Saka, and B. F. Salahu, "An assessment of the roles of irrigation farming in the millennium development goals," African Journal of Agricultural Research, vol. 4, no. 5, pp. 445–450, 2009.

5. D. Crevoisier, Z. Popova, J. C. Mailhol, and P. Ruelle, "Assessment and simulation of water and nitrogen transfer under furrow irrigation," Agricultural Water Management, vol. 95, no. 4, pp. 354–366, 2008.

6. T. J. Trout, "Furrow irrigation erosion and sedimentation: on-field distribution," Transactions of the American Society of Agricultural Engineers, vol. 39, no. 5, pp. 1717–1723, 1996.

7. R. E. Sojka, D. L. Bjorneberg, T. J. Trout, T. S. Strelkoff, and M. A. Nearing, "The importance and challenge of modeling irrigation-induced erosion," Journal of Soil and Water Conservation, vol. 62, no. 3, pp. 153–162, 2007.

8. J. K. Adewumi, "Soil loss tolerance limit for effective conservation measure in Samaru, Northern Nigeria," Savanna Journal of Agricultural Mechanization, vol. 2, no. 1, pp. 23–27, 2000.

9. J. K. Adewumi, O. J. Mudiare, and J. J. Owonubi, "Soil erosion prediction in Samaru-Zaria: comparison of two predictive models," Savanna Journal of Agricultural Mechanization, vol. 3, no. 1, pp. 9–13, 2001.

10. A. C. Odunze, "Mulching practice in a semi-arid zone of Nigeria for soil erosion control and grain yield of maize," Journal of Sustainable Agriculture, vol. 20, no. 2, pp. 31–40, 2002.

11. B. Bakhtiari and A. M. Liaghat, "Seasonal sensitivity analysis for climatic variables of ASCE-Penman-Monteith model in a semi-arid climate," Journal of Agricultural Science and Technology, vol. 13, pp. 1135–1145, 2011.

12. B. G. Leib, C. A. Redulla, R. G. Stevens, G. R. Matthews, and D. A. Strausz, "Erosion control practices integrated with polyacrylamide to reduce sediment loss in furrow irrigation," Applied Engineering in Agriculture, vol. 21, no. 4, pp. 595–603, 2005.

13. T. J. Trout and W. H. Neibling, "Erosion and sedimentation processes on irrigated fields," Journal of Irrigation & Drainage Engineering—ASCE, vol. 119, no. 6, pp. 947–963, 1993.

14. O. Nicetic, A. Lugg, P. H. Sen, L. T. H. Nga, and E. Fliert, "Farmers perception of soil erosion as a risk to their livelihood scenario analysis with farmers in the northern mountainous region of Vietnam," inProceedings of the 3rd International Conference on Conservation Agriculture in Southeast Asia, p. 362, 2012.

15. A. E. Ajayi and I. M. F. Horta, "The effect of spatial variability of soil hydraulic properties on surface runoff process," in Anais XIII Simpósio Brasileiro de Sensoriamento Remoto, pp. 3243–3248, Florianópolis , Brazil, 2007, 21-26 INPE.

16. G. O. Schwab, D. P. Fangmeier, W. J. Elliot, and R. K. Frevert, Soil and Water Conservation Engineering, John Wiley & Sons, New York, NY, USA, 4th edition, 1993.

17. G. J. Wall, W. T. Dickinson, R. P. Rudra, and D. R. Coote, "Seasonal soil erodibility variation in southwestern Ontario," Canadian Journal of Soil Science, vol. 68, no. 2, pp. 417–424, 1988.

18. D. L. Carter, "Furrow irrigation erosion lowers soil productivity," Journal of Irrigation & Drainage Engineering, vol. 119, no. 6, pp. 964–974, 1993.

19. M. J. Brown, D. L. Carter, G. A. Lehrsch, and R. E. Sojka, "Seasonal trends in furrow irrigation erosion in southern Idaho," Soil Technology, vol. 8, no. 2, pp. 119–126, 1995.

20. D. L. Bjorneberg and R. E. Sojka, "Effects of a new polysaccharide-based amendment on furrow irrigation infiltration and erosion," Transactions of the ASABE, vol. 51, no. 2, pp. 529–534, 2008.

21. P. Krause, D. P. Boyle, and F. Bäse, "Comparison of different efficiency criteria for hydrological model assessment," Advances in Geosciences, vol. 5, pp. 89–97, 2005.

22. D. N. Moriasi, J. G. Arnold, M. W. Van Liew, R. L. Bingner, R. D. Harmel, and T. L. Veith, "Model evaluation guidelines for systematic quantification of accuracy in watershed simulations," Transactions of the ASABE, vol. 50, no. 3, pp. 885–900, 2007.

23. K. G. Renard, G. R. Foster, G. A. Weesies, D. K. McCool, and D. C. Yoder, Predicting Soil Erosion by Water: A Guide to Conservation Planning with the Revised Universal Soil Loss Equation, US Department of Agriculture, Agricultural Research Services, Agricultural Handbook 703, 1996.

24. A. P. Savva and K. Frenken, Irrigation Manual Planning, Development Monitoring and Evaluation of Irrigated Agriculture with Farmer Participation Volume I Modules 1–6. SAFR/AGLW/DOC/006 Food and Agriculture Organization of the United Nations (FAO) Sub-Regional, Sub-Regional Office for East and Southern Africa (SAFR), Harare, Zimbabwe, 2002.

25. W. R. Walker, "Guidelines for designing and evaluating surface irrigation systems," FAO Irrigation and Drainage Paper, 1989.

# Chapter 10

# Soil Hydrological Attributes of an Integrated Crop-Livestock Agroecosystem: Increased Adaptation through Resistance to Soil Change

**Mark A. Liebig,[1] Don L. Tanaka,[1] Scott L. Kronberg,[1] Eric J. Scholljegerdes,[2] and Jim F. Karn[1]**

[1]Northern Great Plains Research Laboratory, USDA-ARS, P.O. Box 459, Mandan, ND 58554-0459, USA

[2]Department of Animal and Range Sciences, New Mexico State University, 218 Knox Hall, Las Cruces, NM 88003-8003, USA

## ABSTRACT

Integrated crop-livestock systems have been purported to have significant agronomic and environmental benefits compared to specialized, single-enterprise production systems. However, concerns exist regarding the effect of livestock in integrated systems to cause soil compaction, thereby decreasing infiltration of water into soil. Such concerns are compounded by projections of more frequent high-intensity rainfall events from anticipated climate change, which would act to increase surface runoff and soil erosion. A study was conducted to evaluate the effects of residue management, frequency of hoof traffic, season, and production system (e.g., integrated annual cropping versus perennial grass) on infiltration rates from 2001 through 2008 in central North Dakota, USA. Imposed treatments had no effect on infiltration rate at three, six, and nine years after study establishment, implying that agricultural producers should not be concerned with inhibited infiltration in integrated annual cropping systems, where winter grazing is used. The use of no-till management, coupled with annual freeze/thaw and wet/dry cycles, likely conferred an inherent resistance to change in near-surface soil properties affecting soil hydrological attributes. Accordingly, caution should be exercised in applying these results to other regions or management systems.

# 1. INTRODUCTION

The sustainability of agricultural production in the northern Great Plains of North America is inextricably linked to the amount and frequency of precipitation. Effective use of precipitation by plants requires the retention and movement of water into the soil, which is affected by vegetative cover and soil structure [1]. Management practices that either remove vegetative cover or deteriorate near-surface soil structure slow infiltration rates [2, 3]. Consequences of slower infiltration rates include greater surface runoff, increased erosion, and decreased on-site productivity. Such outcomes may be exacerbated by projected changes in climate, where a more vigorous hydrological cycle is expected to lead to more frequent high-intensity rainfall events [4]. Accordingly, understanding management effects on infiltration rates is important to ascertain potential off-site environmental impacts as well as long-term consequences to agricultural production.

Integrated crop-livestock systems have been purported to improve agricultural productivity, environmental quality, operational efficiency, and economic performance relative to specialized, single-enterprise production systems [5]. Benefits from crop-livestock integration emanate from production synergies brought about by using crops and crop residues for livestock feed while capturing nutrients from livestock wastes for crop production. While many attributes of integrated crop-livestock systems are positive, there are concerns regarding the role of hoof traffic from livestock in these systems to adversely affect near-surface soil conditions, with concomitant impacts on soil hydrological characteristics [6].

In 1999, a team of researchers at the USDA-ARS Northern Great Plains Research Laboratory initiated an integrated crop-livestock study aimed at evaluating production, nutritional, and environmental aspects of overwintering dry bred cows on swathed annual crops [7, 8]. From 2001 through 2008, infiltration rates were measured in the spring and fall of each year at select locations in the established treatments, with the overarching purpose of quantifying treatment effects on near-surface soil hydrological attributes. Specifically, objectives of the research effort were to evaluate effects of residue management, frequency of hoof traffic, season, and production system (e.g., integrated annual cropping versus perennial grass) on infiltration rates for an integrated crop-livestock study located in central North Dakota, USA. As a guiding hypothesis for this research effort, we postulated that perennial grass pastures could be converted to a winter-grazed annual cropping system without adversely affecting soil hydrological attributes.

# 2. MATERIALS AND METHODS

## 2.1. Site and Treatment Description

The research site was located at the USDA-ARS Northern Great Plains Research Laboratory southern research station approximately 5 km south of Mandan, North Dakota, USA (lat. 46°46′35″, long. 100°54′20″). Soils at the site were Temvik-Wilton silt loams (FAO: Calcic Siltic Chernozems; USDA: Fine-silty, mixed, superactive, frigid Typic and Pachic Haplustolls) on a gently rolling upland (approximately 2% slope). Both Temvik and Wilton soils are considered well drained based on USDA-NRCS drainage class characterizations [9].

In 1999, two 6.0 ha crested wheatgrass (Agropyron desertorum (Fisch. ex. Link) Schult.) pastures were sprayed twice with glyphosate (N-(phosphonomethyl) glycine) in mid-May at 0.7 kg a.i. $ha^{-1}$ and converted to an annual crop sequence of oat/pea (Avena sativa L./Pisum sativum L.), triticale/sweet clover (Triticum aestivum ×Secale cereale/Melilotus officinalis L.), and corn (Zea mays L.) using no-till planting techniques. Beginning in 2007, the crop sequence was amended to oat/alfalfa (Medicago spp.)/hairy vetch (Vicia villosa Roth)/red clover (Trifolium pratense L.), Brown midrib sorghum-sudangrass (Sorghum bicolor L. Moench)/sweet clover/red clover, and corn. Each phase of the crop sequence was present each year in both pastures, which served as replicates. Oat and triticale/sorghum crop mixtures were planted in mid- to late May, while corn was planted in early June. All crops were planted with a John Deere 750 no-till drill in 19 cm rows (John Deere, Moline, Ill).

Fertilizer was applied during planting every year in all crop phases as urea (70 kg N ha$^{-1}$ for all crops from 1999 to 2006, 35 kg N ha$^{-1}$ for oat and sorghum crop mixtures, and 70 kg N ha$^{-1}$ for corn in 2007 and 2008) and monoammonium phosphate (6 kg N ha$^{-1}$ and 11 kg P ha$^{-1}$ for all crops every year). Weeds were managed using preemergent herbicides following recommended practices by local farmers.

The oat and triticale/sorghum crop mixtures were harvested for grain between mid-August and early September with the straw spreader removed from the combine, which created a swath of straw and chaff for winter grazing. The corn was swathed for forage in mid- to late September. Crop swaths were oriented parallel to the direction of each crop or crop mixture (e.g., oat, triticale/sorghum, corn). Each crop or crop mixture was split into three residue management treatments including no residue removal (CONTROL), residue removal with a baler (HAYED), and residue removal by grazing with livestock (GRAZED) (Figure 1). The CONTROL and HAYED treatments were randomly assigned within each crop or crop mixture at one end of the plot area, whereas the GRAZED treatment extended across the majority of the plot area and could not be randomly assigned due to logistical considerations associated with field operations and livestock management. Plot area within a crop or crop mixture was 0.05 ha for the CONTROL and HAYED treatments and 1.69 ha for the GRAZED treatment.

The swathed crop residues from the cropping sequence in the GRAZED treatment represented winter forage for ten 4–6-year-old dry bred Hereford cows, due to calve in late March. Grazing commenced in mid-November and ended in mid-February, with the oat crop mixture grazed first, triticale/sorghum crop mixture second, and corn last. Access to crop swaths was controlled by an electric fence and portable fence posts, which allowed for grazing on one crop or crop mixture at a time. Portable fences were oriented at right angles to the swaths within a crop or crop mixture and were moved daily to provide fresh forage for livestock. A shelter and water fountain was located on the end of each pasture within the GRAZED treatment. For each crop or crop mixture, grazing was initiated closest to the shelter and water fountain and extended toward CONTROL and HAYED treatments over time. Accordingly, the direction of grazing within a crop or crop mixture contributed to greater livestock traffic closer to the shelter and water fountain (Clay Erickson, personal communication).

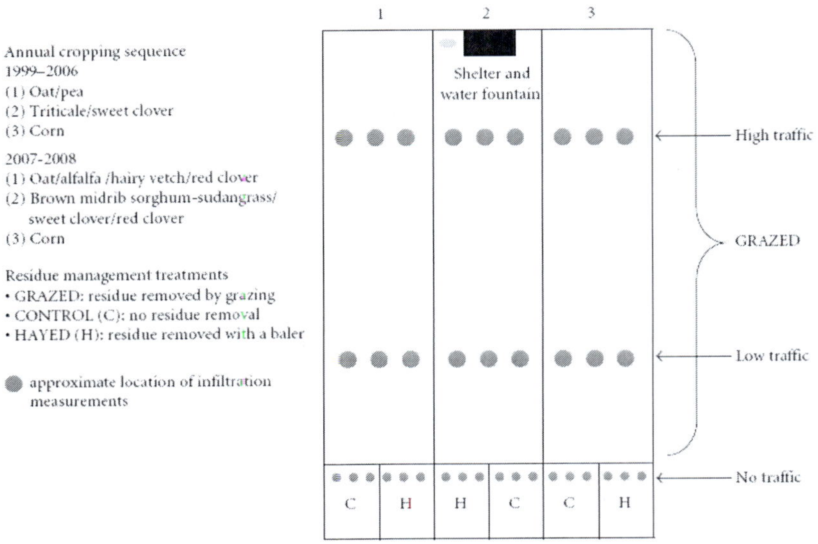


**Figure 1.** Diagram of cropping sequence, residue management treatments, and hoof-traffic zones within an integrated annual cropping pasture near Mandan, ND.

Two 6.0 ha western wheatgrass (Pascopyrum smithii (Rydb.) Love) pastures were used as a perennial grass treatment for comparison of infiltration results with the annually cropped treatments. Swathed grass from the pastures was used as winter forage from 1999 through 2002 for the same number of cows managed similarly as outlined above. Due to a lack of adequate forage due to drought in 2002 and 2003, the perennial grass treatment was hayed but not grazed in 2003. From 2004 through 2008, the perennial grass treatment was lightly grazed with ten Hereford or Angus cows from mid-October to mid-January. Readers are encouraged to review previous reports for additional details regarding crop and livestock management related to the study [7, 8].

## 2.2. Infiltration Measurements

Infiltration measurements were made using single-ring infiltrometers [10]. Briefly, measurements were made by inserting an aluminum ring (i.e., a 15 cm i.d. irrigation pipe cut to a 15 cm length) into the soil to a 7.5 cm depth and applying two separate applications of water within the enclosed space of the ring. The volume of water for each application was equivalent to a 2.54 cm depth (1 inch) within the ring. The time necessary for each application of water to infiltrate into the soil was recorded using a stopwatch. To adjust for potential differences in antecedent water content among treatments, only data from the second water application were used.

Infiltration measurements were made during the spring (April) and fall (October) of 2001/2002, 2004/2005, and 2007/2008. The timing of measurements corresponded to the time after crops had been swathed but not grazed and after the swathed crops had been grazed but not replanted. Measurements in the CONTROL and HAYED treatments were made using nine infiltrometers, oriented randomly in each treatment but between crop rows. Measurements in the GRAZED treatment were made in two transects differing in frequency of hoof traffic and also between crop rows. Nine infiltrometers (three per crop strip) were placed in a transect perpendicular to the swaths approximately 100 m from the shelter and water fountain (representing high-traffic (HT)). A similar transect was established approximately 200 m from the shelter and water fountain (representing low-traffic (LT)). Measurements in the western wheatgrass pastures mirrored that in the annual crop treatments, but with fewer infiltrometers (three in the CONTROL, HAYED, and GRAZED hoof traffic transects).

Three data transformations were evaluated prior to use of untransformed data for statistical analyses. Distribution of log, square root, inverse, and untransformed data was evaluated in SAS using the Shapiro-Wilk test to assess normality given the limited number of observations $(n = 207)$ [11]. All four sets of data yielded outcomes significant at $P \leq .001$, implying that the data were not normally distributed.

Effects of residue management, frequency of hoof traffic, season, and production system on infiltration rate were evaluated three, six, and nine years after study establishment with PROC mixed in SAS using a significant criterion of $P \leq .05$ [12]. Tested effects were considered fixed, while replicates and their interaction were considered random. Treatment means were calculated across crop phases. Given that the GRAZE residue management treatment was not randomly assigned in each replicate, caution was exercised when interpreting results.

To facilitate meaningful comparisons, only data from spring infiltration measurements were used unless stated otherwise. Additionally, for frequency of hoof traffic, data from the CONTROL and HAYED treatments were combined to represent a no-traffic treatment (NT). Similarly, only data from the GRAZED treatment were used to compare seasonal effects (i.e., Fall versus Spring) on infiltration rate.

## 3. RESULTS AND DISCUSSION

Livestock integration of an annual crop sequence (GRAZED) did not significantly affect infiltration rate as compared to residue management not involving livestock (HAYED) or residue left in place (CONTROL) (Figure 2(a)). Numerically, however, infiltration rates within the GRAZED treatment were consistently lower among the three residue management treatments at three, six, and nine years after study establishment. Despite the numerically lower infiltration rates in the GRAZED treatment (e.g., approximately 5 cm $hr^{-1}$ in 2002; Figure 2(a)), concerns with potential runoff were negligible at this site as the occurrence of a 5 cm rain over a 1 hr duration is expected to happen at a historical frequency from 10 to 25 yr [13]. Both low- and high-frequency hoof traffic zones (LT, HT) possessed numerically lower infiltration rates as compared to the zone where hoof traffic was excluded (NT) (Figure 2(b)); however, differences among hoof traffic zones were not significant in any year. Moreover, infiltration rates within the GRAZED treatment did not differ between fall and spring in 2001/2002, 2004/2005, or 2007/2008 (Figure 2(c)).

Infiltration rates did not differ between integrated annual cropping and perennial grass grazing systems during any of the measurement times (Table 1). Accordingly, the guiding hypothesis for the study was not disproved, suggesting that perennial grass pastures may be converted to a winter-grazed annual cropping system without compromising infiltration of water into soil.

Maintenance of key hydrological functions in agroecosystems is an essential attribute affecting agricultural productivity and environmental quality. Climate-induced shifts toward more frequent high-intensity rainfall events pose significant challenges to the hydrological functions of agroecosystems and, in turn, agricultural sustainability [14]. As such, anticipated changes in climate may favor agroecosystems that can effectively capture and store significant amounts of water in soil following intense precipitation events. Findings from this study indicate that integrated annual cropping systems utilizing winter grazing may be suitably adapted to a more vigorous hydrological cycle from the standpoint of water capture and storage. Such adaptation emanates from the capacity of the management system to withstand significant changes in near-surface soil physical properties known to affect water infiltration into soil during, and following, conversion from perennial grass. One example of resistance to change in soil physical condition was expressed through the magnitude of treatment effects on soil bulk density, a general indicator of soil compaction. Within the surface 7.5 cm of soil, soil bulk density never exceeded 1.20 Mg m$^{-3}$ over the course of the study nor were soil bulk density values correlated with infiltration rates ($r = -0.07$; $P = .3033$; $n = 207$) (data not shown) [15].

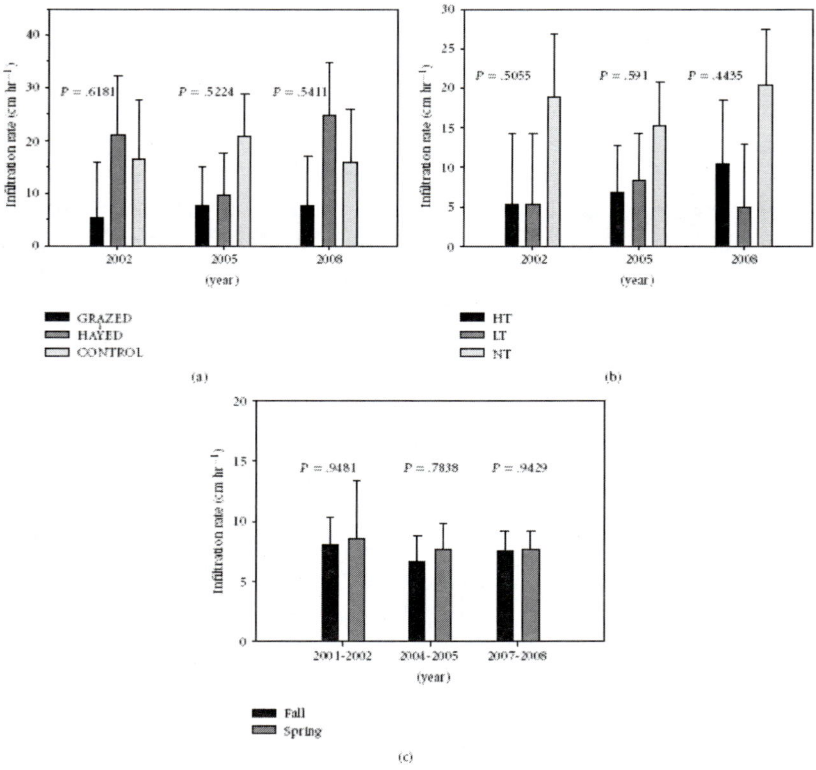


**Figure 2.** Effects of grazing and residue management (a), degree of hoof traffic (b) (HT: high traffic, LT: low traffic, and NT: no traffic), and season (c) on infiltration rate for an integrated crop-livestock system near Mandan, ND. Spring infiltration measurements were used in (a) and (b). Error bars represent ± 1 standard error.

**Table 1.** Mean infiltration rate in the spring of 2002, 2005, and 2008 for integrated annual cropping and perennial grass near Mandan, ND.

| | Infiltration rate (cm $hr^{-1}$) | | |
|---|---|---|---|
| Year | Integrated annual cropping | Perennial grass | *P* value |
| 2002 | 13.0 (5.8)* | 7.1 (7.4) | .5984 |
| 2005 | 10.9 (5.6) | 21.1 (7.1) | .3821 |
| 2008 | 14.0 (3.8) | 5.1 (4.3) | .2649 |

* Values in parentheses represent standard error of the mean.

Pragmatically, results from this study suggest that agricultural producers should not be concerned with infiltration problems in integrated annual cropping systems, where winter grazing is used. Had cattle caused serious soil compaction during winter grazing, it would have been expressed through significantly lower infiltration rates. A combination of management-, climatic-, and edaphic-related factors was likely responsible for the observed resistance in soil change to the treatment perturbations.

No-till conversion of the crested wheatgrass pastures to annual cropping in 1999 may have been critical in conferring an inherent resistance against changes in near-surface soil physical properties caused by subsequent perturbations from cropping and grazing. An absence of tillage during conversion allowed for soil physical properties (soil structure, pore continuity, and pore size) to remain unchanged. This consistency in soil physical condition during conversion was likely critical in buffering soil changes potentially caused by differences in plant type (annual crops versus perennial grass) or implementation of field activities (e.g., spraying, planting, and swathing) over the course of the study.

Utilization of winter grazing in this study likely mitigated potential soil compaction in the GRAZED treatments as the soil is typically frozen from November through March. In the absence of mid-winter thawing periods, soils remain frozen and less susceptible to compaction by hoof traffic. Moreover, soils in the northern Great Plains undergo annual freeze/thaw and wet/dry cycles, which act to fracture compacted soil zones through the expansion/contraction of soil minerals and water in soil pores. Such natural amelioration processes have been found to buffer effects of grazing on near-surface soil compaction in the northern Great Plains [16]. Significant contributions from earthworm tunneling, as observed in the integrated annual cropping system (Jason Gross, personal communication), can also limit effects of compaction on soil water properties through the creation of large biopores.

## 4. CONCLUSIONS

Infiltration of water into soil was found not to be significantly affected by management perturbations within an integrated annual cropping system, where winter grazing was used. The consistent freeze/thaw and wet/dry cycles characteristic of the northern Great Plains, coupled with the use of no-till management, likely played an important role in the outcome of the results. Accordingly, caution should be exercised in applying results from this study to other regions or management systems. Outcomes from the study are nonetheless valuable from the standpoint highlighting the importance of agroecosystem management to minimize changes in soil hydrological functions. Such an attribute may be critical in adapting to a more vigorous hydrological cycle from anticipated climate change.

## ACKNOWLEDGMENTS

The authors gratefully acknowledge the contributions of Justin Hartel, Marvin Hatzenbuhler, and Robert Kolberg for planting and swathing crops, Clay Erickson, Gordy Jensen, and Curt Klein for managing cattle, and John Bullinger, Jason Gross, and Holly Johnson for assisting with infiltration measurements. The US Department of Agriculture (USDA) prohibits discrimination in all its programs and activities on the basis of race, color, national origin, age, disability, and, where applicable, sex, marital status, familial status, parental status, religion, sexual orientation, genetic information, political beliefs, reprisal, or because all or part of an individual's income is derived from any public assistance program (not all prohibited bases apply to all programs). Persons with disabilities who require alternative means for communication of program information (Braille, large print, audiotape, etc.) should contact USDA's TARGET Center at (202) 720-2600 (voice and TDD). To file a complaint of discrimination, write to USDA, Director, Office of Civil Rights, 1400 Independence Avenue, S.W., Washington, DC 20250-9410, or call (800) 795-3272 (voice) or (202) 720-6382 (TDD). USDA is an equal opportunity provider and employer.

Mention of trade names or commercial products in this paper is solely for the purpose of providing specific information and does not imply recommendation or endorsement by the US Department of Agriculture.

## REFERENCES


1. N. C. Brady and R. R. Weil, The Nature and Properties of Soils, Prentice Hall, Upper Saddle River, NJ, USA, 12th edition, 1999.

2. J. L. Pikul and J. K. Aase, "Infiltration and soil properties as affected by annual cropping in the northern Great Plains," Agronomy Journal, vol. 87, no. 4, pp. 656–662, 1995.

3. S. D. Warren, T. L. Thurow, W. H. Blackburn, and N. E. Garza, "The influence of livestock trampling under intensive rotation grazing on soil hydrologic characteristics," The Journal of Range Management, vol. 39, no. 6, pp. 491–495, 1986.

4. M. A. Nearing, F. F. Pruski, and M. R. O'Neal, "Expected climate change impacts on soil erosion rates: a review," Journal of Soil and Water Conservation, vol. 59, no. 1, pp. 43–50, 2004.

5. M. P. Russelle, M. H. Entz, and A. J. Franzluebbers, "Reconsidering integrated crop-livestock systems in north America," Agronomy Journal, vol. 99, no. 2, pp. 325–334, 2007.

6. R. M. Sulc and B. F. Tracy, "Integrated crop-livestock systems in the U.S. corn belt," Agronomy Journal, vol. 99, no. 2, pp. 335–345, 2007.

7. D. L. Tanaka, J. F. Karn, M. A. Liebig, S. L. Kronberg, and J. D. Hanson, "An integrated approach to crop/livestock systems: forage and grain production for swath grazing," Renewable Agriculture and Food Systems, vol. 20, no. 4, pp. 223–231, 2005.

8. J. F. Karn, D. L. Tanaka, M. A. Liebig, R. E. Ries, S. L. Kronberg, and J. D. Hanson, "An integrated approach to crop/livestock systems: wintering beef cows on swathed crops," Renewable Agriculture and Food Systems, vol. 20, no. 4, pp. 232–242, 2005.

9. Soil Survey Staff, “Natural Resources Conservation Service, United States Department of Agriculture. Web Soil Survey,” 2011, http://websoilsurvey .nrcs.usda.gov.

10. M. Sarrantonio, J. W. Doran, M. A. Liebig, and J. J. Halvorson, “On-farm assessment of soil quality and health,” in Methods for Assessing Soil Quality, J. W. Doran and A. J. Jones, Eds., pp. 83–105, Soil Science Society of America, Madison, Wis, USA, 1996.

11. SAS Institute, SAS/STAT User's Guide. Version 6.0, vol. 2, SAS Institute, Cary, NC, USA, 4th edition, 1990.

12. R. C. Littell, G. A. Milliken, W. W. Stroup, and R. D. Wolfinger, SAS System for Mixed Models, SAS Institute, Cary, NC, USA, 1996.

13. D. M. Hershfield, “Rainfall frequency atlas of the United States,” in Weather Bureau technical Paper, vol. 40, US Government Printing Office, Washington, DC, USA, 1961.

14. W. M. Edwards and L. B. Owens, “Large storm effects on total soil erosion,” Journal of Soil and Water Conservation, vol. 46, no. 1, pp. 75–78, 1991.

15. M. A. Liebig, D. L. Tanaka, S. L. Kronberg, E. J. Scholljegerdes, and J. F. Karn, “Integrated crops and livestock in central North Dakota, USA: agroecosystem management to buffer soil change,” Renewable Agriculture and Food Systems. In press.

16. E. Mapfumo, D. S. Chanasyk, M. A. Naeth, and V. S. Baron, “Soil compaction under grazing of annual and perennial forages,” Canadian Journal of Soil Science, vol. 79, no. 1, pp. 191–199, 1999.

# Chapter 11

# Managing Soil Erosion in Northern Ireland: A Review of Past and Present Approaches

**Donal Mullan**

School of Natural Sciences and Psychology, Liverpool John Moores University, Liverpool L3 3AF, England, UK

## ABSTRACT

In recent decades, the off-site transport of sediment and nutrients from agricultural land into the neighbouring natural and built environment has become a more pressing environmental sustainability issue than the on-site threats of soil erosion in many of the world's temperate regions. In the temperate region of Northern Ireland, recent studies have highlighted the off-site issue of soil erosion by water in the present day, and projected that the problem may become more widespread and serious in a changing climate. This review paper examines how this problem is being managed in the present day, and draws on examples of policy in other countries to consider how the role of policy needs to be modified for more effective management. Farmers are generally not adhering to present-day policy and "keeping their land in good agricultural and environmental condition". A range of suggested changes in policy and practice is offered, ranging from educating farmers on erosion mitigation and remediation to developing specific policies aimed at targeting soil erosion and conservation as their sole objective. An increase in the evidence base from measured erosion rates in the field is postulated to be the most likely route to achieving policy changes.

## KEYWORDS

soil erosion; muddy floods; off-site impacts; policy; farmers; agriculture

## 1. INTRODUCTION

Although a natural process which has been operating on landscapes throughout the world since the dawn of settled agriculture [1,2], soil erosion has been greatly accelerated by human activity owing to the clearance of natural ecosystems into agricultural land and the associated working of that land [3]. This has led to a reduction in soil quality and overall agricultural productivity [1]. Termed the "on-site" impacts of soil erosion, issues include reduced water and nutrient-holding capacity of the soil [4,5]; reduced organic matter content [6]; and reduced soil depth to support roots and organisms [7,8]. The "off-site" impacts, meanwhile, consider the fate of soil when particles leave the agricultural land. Adverse off-site impacts include the economic consequences when large quantities of suspended sediment are transported from fields and into homes, settlements and infrastructure in "muddy floods" [9], and the environmental damage associated with sedimentation of sand and gravel-bedded rivers [10] and the adsorption of chemicals onto soil particles and eutrophication of water bodies [2]. A more detailed account of the on-site and off-site impacts of soil erosion can be found in [11].

In the temperate region of Northern Ireland, recent studies [11,12,13,14,15] have highlighted the present-day off-site problem of soil erosion, and have projected that soil erosion could become a more serious issue under the impacts of a changing climate. The most direct of these impacts are changes in the erosive power of rainfall [16,17,18,19,20]. With an increase in the moisture-holding capacity of the atmosphere at a rate of *ca.* 7% per 1 °C rise in temperature (*sensu* the Clausius-Clapeyron relationship), a more vigorous hydrological cycle is created, modifying precipitation characteristics such as amount, frequency, duration, intensity, and their extremes [21]. Indirectly, increases in temperature and $CO_2$concentrations can drive changes in plant biomass, with increasing erosion rates possible due to faster residue decomposition and increased microbial activity [22], and decreasing erosion rates possible owing to an increased soil surface canopy and biological ground cover [23]. In addition, farmers may wish to take advantage of new climatic conditions by adapting the timing of their agricultural operations, with changing planting and harvesting dates and the implementation of new crops all factors with the potential to considerably alter the rates and patterns of soil erosion [22].

Given the present-day problem and future potential for a greater soil erosion problem in Northern Ireland, it is vital that policy is in place to reduce the harmful off-site effects. A review of soil erosion policies has been documented for many countries, including Belgium [24]; Czech Republic [25]; Germany [26]; Norway [27]; Spain [28]; and United Kingdom (England and Wales only) [29,30]. National erosion policies are generally lacking, but EU subsidies and voluntary agri-environmental schemes indirectly have consequences for soil erosion control. The Soil Thematic Strategy [31] is a European Soil Policy which works to ensure sustainable use of the soils of Europe through preventing further soil degradation and preserving its functions, as well as restoring degraded soils to a level of functionality consistent with its intended use [31]. Among the four pillars of the Thematic Strategy is integration with the Common Agricultural Policy (CAP), which has an important role to play in soil management [32]. The EU CAP [33] indirectly requires farmers to exercise erosion control as they must keep their land in "Good Agricultural and Environmental Condition" (GAEC) to qualify for subsidy payments. GAEC guidelines [34] include stipulations on land management between harvesting and planting, such as planting of cover crops or leaving stubble of the harvested crop on the land [34]. The Water Framework Directive (WFD) is an EU directive which, like the CAP, possesses an indirect requirement on farmers to control erosion and its off-site impacts to maintain "good ecological status of surface waters" [35]. In individual EU countries, agri-environmental schemes are also in operation, which can provide additional indirect incentives to farmers to control erosion from their land. Soil erosion in Northern Ireland is indirectly managed by many of these EU and national directives and agri-environmental schemes, with guidelines on farmer practices from commissioned reports also acting as an avenue for encouraging (but not enforcing or even incentivizing) management. The report "Climate Change, land management and erosion in the organic and organo-mineral soils in Scotland and Northern Ireland" [36] included information on present and future drivers of soil erosion, and offered advice on methods for erosion reduction, such as confining sheep grazing in erosion-prone areas to the growing season. This report is the only government-funded commissioned piece of work which relates to soil erosion and erosion management issues in Northern Ireland.

The aim of this paper is to review policies directly or indirectly aimed at reducing soil erosion and its off-site impacts in Northern Ireland, and evaluate the effectiveness of these policies for the past and present day. Drawing on evidence of successful approaches to soil erosion policy in other countries, the paper will then consider how policies could be revised for effective management of the soil erosion problem in the future.

# 2. APPROACHES AND BARRIERS TO MANAGING SOIL EROSION IN NORTHERN IRELAND

## 2.1. Past and Present-Day Policies

Since agricultural and more recently "agri-environmental" policies dictate subsidy payments to farmers for agricultural production and environmental awareness, they are fundamental to managing the on-site and off-site impacts of soil erosion. In Europe, two contrasting policy phases are considered, with each bringing about a different response of soil erosion and its associated impacts. These can be divided into (1) agricultural policies encouraging intensive food production post-World War 2; and (2) agri-environmental policies with a more environmental focus following the United Nations Conference on Environment and Development (UNCED) or "Earth Summit" in Rio de Janeiro in 1992.

### 2.1.1. Agricultural Policies Post-World War 2

The most significant agricultural policies affecting the on-site and off-site impacts of soil erosion in the UK (including Northern Ireland) arose from the implementation of the European Union (EU) CAP in 1962 and its incorporation into UK agricultural policy following its inception into the EU in 1973. The CAP was established in order to boost European food production in the face of post-war shortages and increasing farm incomes, supported by subsidy payments to farmers for agricultural production [37]. Accompanying this drive towards agricultural intensification were a number of changes which began through the earlier Agriculture Act (1947) in the UK. Such changes included the application of more fertilisers [38]; higher yielding cereal varieties [39]; the introduction of pesticide use; use of bigger and heavier machinery; the removal of hedgerows and field boundaries [40]; a decrease in the number of holdings and workers; an increase in farm size [41]; and a large increase in winter cereals by a factor of about three, particularly in England and Wales [42]. All such practices were designed to facilitate more efficient agriculture and higher agricultural productivity per hectare, but had the added effect of markedly increasing the vulnerability of the land to runoff and erosion across England and Wales [43]. Although only limited measured data exist to assess the problem in the study area, past evidence for soil erosion in Ireland [12,14,44] indicates that the drive towards agricultural productivity had the effect of resulting in widespread erosion across many parts of the island.

## 2.1.2. Towards Agri-Environmental Policies

With the growing recognition that agricultural intensification was affecting not only soils but also the wider environment (e.g., [7]), reforms to the CAP since the late 1980s adopted a more environmental focus, including the introduction of voluntary set-aside in 1988 and compulsory set-aside in 1992 to reduce the production of combinable crops [29]. Although not implemented as a direct measure to reduce soil erosion, a number of case studies in England [44,45,46,47,48] documented the success of set-aside on slopes vulnerable to erosion in reducing soil loss and leading to a decline in connectivity of flow from field to field and field to watercourse, thereby also reducing problems associated with the off-site impacts of erosion.

A review of the CAP in 1999 [49] led to greater emphasis on environmental protection and conservation of the countryside in addition to food production. To receive payments via the Single Farm Payment (SFP) scheme (2005), farmers would have to keep their land in good condition and for the first time would have to ensure erosion and runoff from their land was minimised [29]. As part of the SFP, farmers were also encouraged to enter agri-environmental schemes such as the Countryside Management Scheme (CMS) or Environmentally Sensitive Areas (ESAs) in Northern Ireland. Through such schemes, farmers are offered financial incentives to maintain and enhance biodiversity and assist the implementation of the WFD [35]. The latter was incorporated into UK law in 2003 as a piece of legislation to protect and enhance water quality. Under the directive, the chemical and ecological conditions of all inland and coastal waters should not deteriorate from their present state and must be of "good surface water status" by

2015 [35]. Similar to the qualification criteria for receiving subsidy payments through the SFP, the WFD ensures farmers must take account of erosion and runoff from their land, and are responsible for ensuring that off-site impacts of erosion from their land does not compromise "good water surface status" by 2015. The Farm Nutrient Management Scheme (FNMS) was introduced in 2005 to provide farmers with financial incentives to increase their slurry storage capacity, thereby ensuring more efficient use of available nutrients and reducing the runoff of nutrients into surface waters [50]. In this respect, the FNMS also seeks to achieve compliance with the WFD in ensuring good water quality standards in Northern Ireland. A summary of the main agri-environmental policies indirectly affecting soil erosion in Northern Ireland is shown in Table 1.

**Table 1.** Main agri-environmental policies indirectly affecting soil erosion in Northern Ireland and their objectives.

| Year | Policy | Objective |
|---|---|---|
| 2003 | Water Framework Directive | To achieve good ecological and chemical status in surface waters |
| 2005 | Single Farm Payment | To keep land in good agricultural and environmental condition |
| 2007–2013 | Countryside Management Scheme/ Environmentally Sensitive Areas | To improve the environment and the countryside through land management |
| 2005 | Farm Nutrient Management Scheme | To reduce nutrient runoff to surface waters |

## 2.2. Are Present-Day Policies Helping to Reduce Erosion?

### 2.2.1. Policy Effectiveness in Reducing Erosion

As noted in the previous section, a number of studies from England [44,45,46,47,48] have highlighted the success of set-aside of land on slopes vulnerable to erosion. In the early 1990s in the South Downs, England, one particular farmer put some of his land into set-aside since it had been flooding a downslope housing estate since the mid-1980s with sediment from his eroding land. The farmer was able to take this action without loss of subsidy due to the introduction of voluntary set-aside in 1988 and compulsory set-aside in 1992 as part of the CAP. Since set-aside, the housing estate has never been flooded, not even during the winter of

2000/01 when many other estates suffered muddy floods [29]. There are also instances, however, where agri-environmental policies can lead to increased erosion. It has been demonstrated through field observations and modelling studies in Spain that set-aside can lead to increased soil erosion due to the treatment of set-aside land as unseeded fallow [28]. Land is left as unseeded fallow to increase soil moisture and fertility for subsequent rotations due to the semi-arid nature of the Mediterranean climate. With low annual rainfall totals in many parts of Spain, intensive cropping exhausts soil resources of water and nutrients, meaning unseeded fallow is one means of restoring fertility for future crop rotations [28]. Although set-aside in the temperate climate of Ireland is unlikely to result in a similar land management in the present day, climate change projections illustrate future summer drying [51,52], indicating that increased erosion due to the treatment of set-aside land as unseeded fallow could be a possibility in the future. Of course, by this time, set-aside may no longer act as a funded agri-environmental policy due to the changing nature of the policy landscape.

### 2.2.2. Issues of Farmer Compliance

Although the range of policies designed to improve water quality and concern for the wider environment look as though they may provide the answers to reducing the effects of soil erosion, the situation in reality may be more complicated. This is because, according to [29], many farmers in Britain are not complying with recent agri-environmental policies and keeping their land in "good agricultural and environmental condition". Personal observations in Northern Ireland support this, where numerous incidences of erosion have been spotted over a three year period between 2008 and 2011 (Figure 1) [14].

There may be various reasons responsible for lack of compliance. Firstly, since UK policy is designed primarily to reduce the off-site impacts of erosion such as diffuse pollution and muddy floods rather than the on-site conservation of soil, the cost of failure to comply with policy regulations is no greater to the farmer than it is to the general tax-paying public. In Northern Ireland, lack of compliance with mandatory regulations and uptake of voluntary regulations may be a greater problem than it is in Britain. This is because colloquial evidence suggests that farmers in Northern Ireland generally do not consider soil erosion to be a problem. When the father of the author, a local farmer, was asked whether he would consider applying for the CMS, he replied no because he thought the scheme was "eejiting" (Irish slang for foolishness). Whether his views are representative of the farming community in Northern Ireland requires further research, but the general lack of concern amongst farmers regarding the impacts of soil erosion is evident. The second possible reason for lack of compliance with policy regulations is because of the relatively subjective nature of the policy objectives with respect to environmental issues. Statements such as "to achieve good water surface status" and "to keep land in good agricultural and environmental condition" (see Table 1) are open to interpretation and thus easier for farmers to evade and still receive subsidy payments.

# 3. FUTURE POLICY CONSIDERATIONS FOR NORTHERN IRELAND

Given the potential for increased soil erosion under the impacts of future climate change in Northern Ireland (see Figure 2) [11,15], the need for greater enforcement with respect to existing agri-environmental policies and/or the development of a new set of policies developed specifically to tackle soil erosion from agricultural land may be required. In order to be robust and long-term, future policy changes should be (1) based on experience; (2) relevant to a wide range of possible future climatic conditions; and (3) cognisant of the many possible agents of societal change [53].

## 3.1. Clearer Definitions for Environmental Standards

As stated previously, qualitative statements with respect to achieving environmental standards clearly makes it easier for farmers not to fulfil them and still receive subsidy payments. An exception is the EU Fish Directive [54], where guideline values are in place for EU member states to keep suspended solids and nutrient levels in surface waters below certain thresholds, e.g., ≤25 mg/L of suspended solids and ≤0.01 mg/L of nitrites in salmonid waters [55]. If a more rigorous metric was in place, then farmers would be under greater pressure to minimise erosion from their land in order to meet the requirements of their subsidies. In terms of enforcement, the EU may wish to consider implementing a water quality target through its WFD scheme in the same way as the 1 t/ha/year limit set by USDA for the USA [56] to control pollution from sediment and maintain water quality standards. A threshold for on-site problems could also be set, e.g., those set in [56].

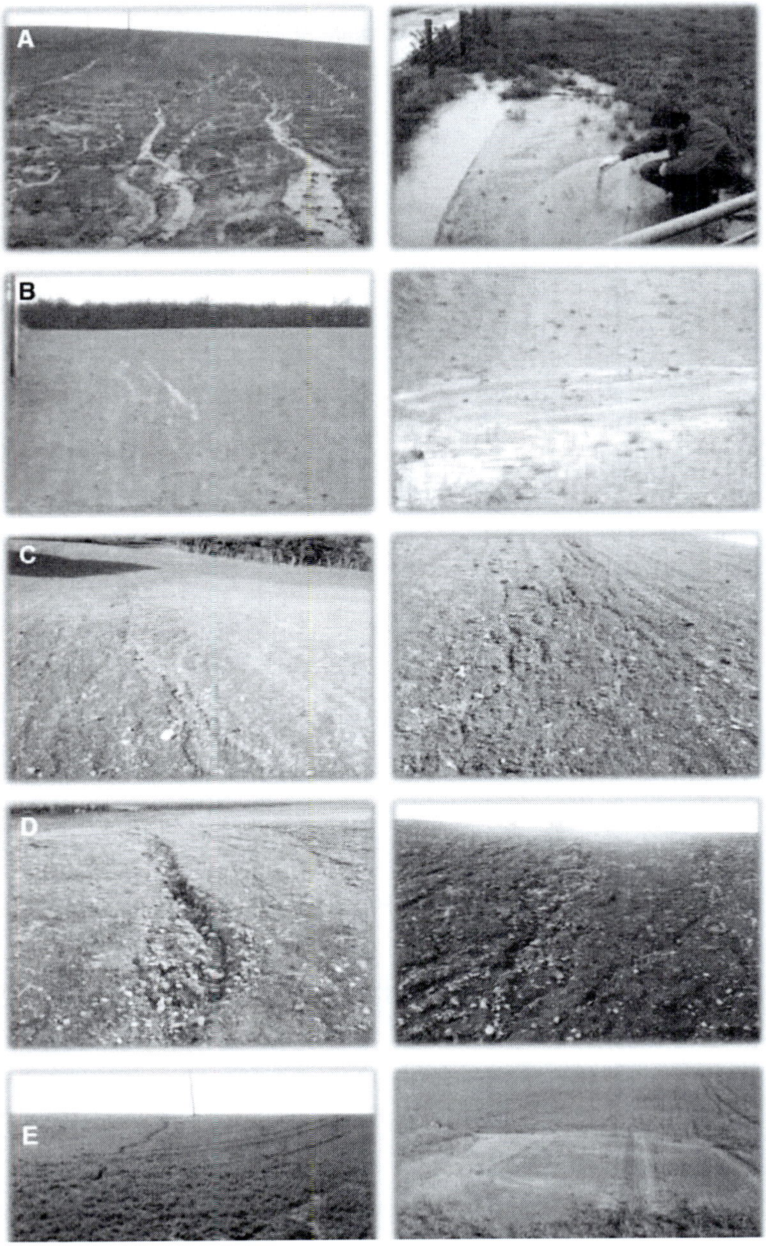


**Figure 1.** Five separate soil erosion events on agricultural land in County Tyrone between 2008 and 2012. (**A**) Loughmuck, Omagh, Co. Tyrone; (**B**) Ballynahatty, Omagh, Co. Tyrone; (**C**) Kevlin, Omagh, Co. Tyrone; (**D**) Aughafad, Clogher Valley, Co. Tyrone; (**E**) Blackfort, Omagh, Co. Tyrone.

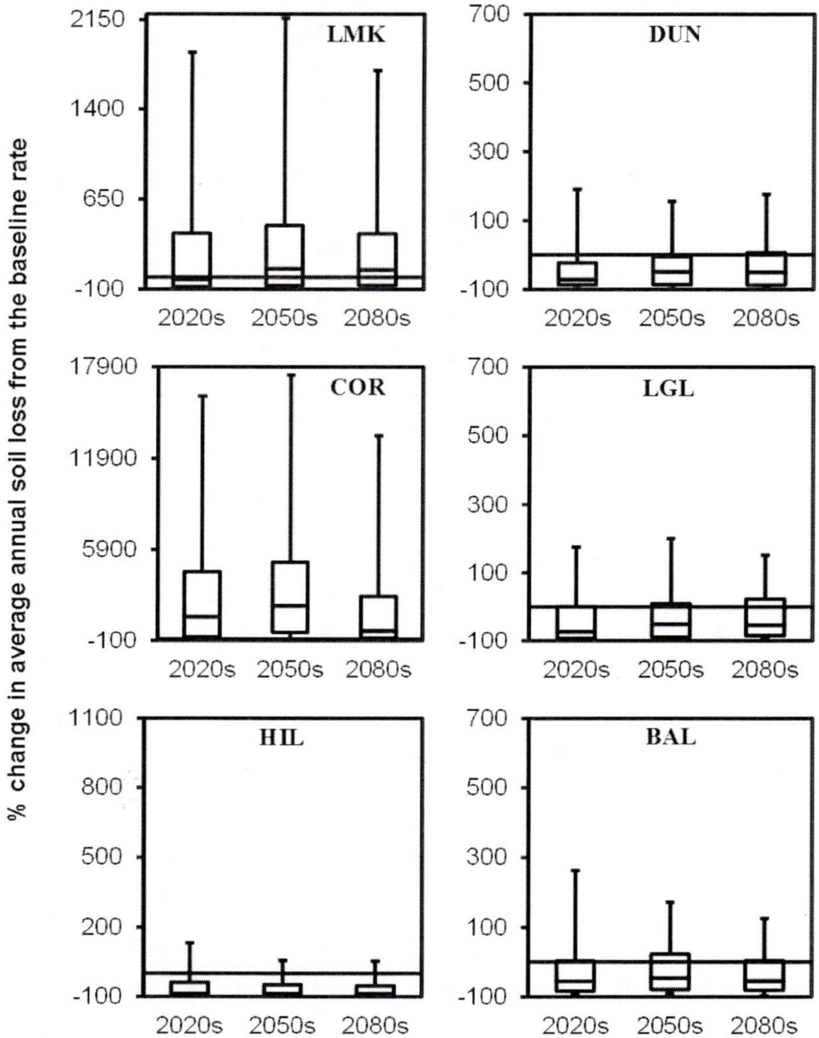


**Figure 2.** Percentage change in soil loss for three future time periods relative to the modelled present-day baseline erosion rate for six hillslopes in Northern Ireland under future climate and land use change scenarios [15]. The full distribution at each hillslope accounts for 216 future scenarios (3 GCMs × 2 emissions scenarios × 3 ensembles × 4 land use scenarios × 3 future time periods). Lower whisker: minimum; lower box line: 25th percentile; middle box line: median; upper box line: 75th percentile; upper whisker: maximum. LMK: Loughmuck; DUN: Dunadry; COR: Corrard; LGL: Loughgall; HIL: Hillsborough; BAL: Ballywalter.

## 3.2. Greater Enforcement of Subsidy Requirements

In addition, the regularity of inspections and penalties for breaching rules as part of any subsidy or agri-environmental scheme may need to be revised to ensure greater land management and environmental responsibility. Under the SFP, for example, the penalty for failure to comply with keeping land in good agricultural and environmental condition is just 3% of the annual subsidy payment [57], which is unlikely to be a considerable enough deterrent to farmers to ensure soil erosion from their land is at the forefront of their consciousness. Consideration may be given to heavier penalties for breaching cross-compliance.

## 3.3. Raising Awareness through Education

The reduction of soil erosion and the runoff of sediment and associated chemicals and fertilisers into nearby water courses is a central objective for "keeping land in good agricultural and environmental condition" as part of the cross-compliance regulations for SFP payments and as part of the conditions for meeting a number of agri-environmental schemes. Despite this, none of these policies make any reference to how individual farmers should address the issue. Up until the inception of the SFP in 2005, awareness of soil erosion and methods to alleviate it were only available to local farmers in the UK through pamphlets distributed by organisations such as the Ministry of Agriculture, and Forestry and Fisheries (MAFF) (now the Department of Environment Food and Rural Affairs: DEFRA) [58,59]. As part of the cross-compliance for SFP subsidy payments from 2005 onwards, the "Guidance for Soil Management" [60] states that farmers in England must "*consider the risk of runoff and erosion when planning what to grow*" and advises that "*where severe erosion occurs, earth banks or other physical barriers may be used as a last resort to check the flow of water and reduce off-site impacts*". This is the only piece of soil conservation advice currently written into UK policy. It is important to note, however, that this applies only to England and is not part of the SFP cross-compliance policy in Northern Ireland. In this respect, whilst policies exist that require farmers to minimise erosion from agricultural land (amongst other objectives) in Northern Ireland, no advice is offered in terms of prevention through effective land management or remedies to the problem when erosion actually occurs.

Education is a critical step in helping to minimise erosion and its off-site impacts. Information needs to be made available to local farmers regarding the recognition of soil erosion in their fields, as well as advice on how to prevent erosion through effective land management and deal with the problem if erosion does occur. An approach similar to the pamphlets developed by MAFF in England in the mid-1980s may be a good starting point to help raise awareness of soil erosion and its adverse impacts amongst farmers in Northern Ireland. Taking this action in the present-day may help alleviate present-day off-site impacts of soil erosion through better land management, whilst also providing an early platform to build recognition amongst farmers with respect to the potential for increased soil erosion.

## 3.4. Implementation of Conservation Strategies

In other EU countries, examples exist whereby conservation strategies are written into agricultural or agri-environmental policy. Taking Flanders, Belgium as a particular case study, as part of the cross-compliance for the Single Payment Scheme (SPS) (equivalent to the SFP in Northern Ireland), farmers whose fields are classified as very susceptible to erosion must control the problem through at least one "farmer's action". Typical farmer's actions include the implementation of cover crops, contour tillage, grass buffer strips and conservation tillage [24]. Given the potential for muddy flooding and other severe off-site effects of erosion under certain future scenarios in Northern Ireland, there may be a case that future policies could include the stipulation that a soil conservation technique is implemented in areas susceptible to erosion, as is the case in present-day Belgium. This would require the development of a soil erosion risk assessment procedure similar to that developed by DEFRA [59] and assessed by [10] for controlling water erosion in England, in order to enable vulnerable areas to be targeted for conservation practices. Indeed the existing DEFRA scheme for England could be directly implemented in Northern Ireland. Any such conservation should fulfil the low-regret, reversible, flexible, low-cost and "soft solutions" approach that targets protecting the soil surface from raindrop impact and runoff by means of permanent vegetation rather than hard engineering [24].

There have also been a number of instances where conservation strategies have been advised in individual legal cases in England, where muddy floods caused disruption to nearby properties. When houses and roads suffered muddy flooding in the winter of 2000/2001 in Suffolk, England originating from a field of outdoor pigs, a map was produced (Evans, unpublished) showing where grass margins could be implemented to protect the most vulnerable slopes and thus reduce runoff and erosion in future years. When this proposal was implemented following threat of legal action, no flooding to properties or major roads occurred [29]. In 2001 a similar case study from an outdoor pig field in Suffolk resulted in a court case settled in favour of the plaintiff following muddy flooding of a property, with the consequence that parts of two fields can no longer be used for rearing pigs [61]. Given the potential for increased muddy flooding in Northern Ireland over the coming century (See Figure 2) [11,15], these sorts of legal cases may become prevalent in the future. Again, this provides further weight to the argument that risk maps would be useful to identify the most susceptible areas to muddy floods to help target mitigation measures. The DEFRA risk mapping criteria for England and Wales could be used. It is acknowledged, of course, that the challenge would then become getting farmers to adopt control measures resulting from this mapping.

## 3.5. Specific Agri-Environmental Measures

Again, looking towards Flanders, Belgium as a notable case study, there exist agri-environmental schemes designed to tackle soil erosion as their primary objective as part of the Flemish Rural Development Programme, driven by local authorities and residents of flooded villages [38]. The crucial element with this example is the financing model used in Flanders. Since the Erosion Decree in 2001, municipalities in the Flemish region became eligible for subsidies to draw up an erosion mitigation scheme on a voluntary basis [62]. Subsidies for farmers and land-owners are financed 75% by the Flemish government, and 25% by the municipality, whilst construction costs of mitigation measures are financed 75% by the government, 15% by the province, and 10% by the municipality [63]. These incentive-based measures are well accepted and adopted by farmers in the Flanders region. Personal visits from farm advisers (partly paid by the government) to facilitate the delivery of policy measures to farmers are heralded as the formula for success in the adoption of measures by farmers. As a result of this personal approach, farmers are more confident, cooperative and know who to contact for questions, as demonstrated by interview responses from Flemish farmers [24]. Lessons could be learned from this case study for application to other countries where soil erosion is a problem, including Northern Ireland. Voluntary agri-environmental schemes in Northern Ireland are also available through the Northern Ireland Rural Development Programme. The Northern Ireland Countryside Management Scheme [64] is one such scheme, but is aimed predominantly at enhancing biodiversity, and does not provide funding for measures to reduce soil erosion and its impacts. An erosion policy at a national level is an important means to gain institutional memory in managing the problem of soil erosion and muddy floods [65]. Government support of funding in the form of an erosion-specific agri-environmental scheme to help implement conservation measures could be greatly effective in reducing erosion in high-risk areas. According to [66], however, the current range of EU policies, schemes and directives has the potential to address all recognised soil degradation processes throughout the EU, without the need for specific schemes.

## 3.6. Research Required to Facilitate Suggested Approaches

Policy must, however, if it is to have some evidence base, lag behind research [29]. As demonstrated in Britain, it may take some time to persuade policy makers that there is a problem which needs to be addressed. Monitoring of erosion and runoff are critical to enable appropriate policy making, not only to know the severity of present-day erosion but also as a basis to help build and validate models to project future erosion rates to help inform future policy. In the flow diagram representing these phases in Figure 3, the situation with regards to Northern Ireland is currently only at the "recognition" stage. Some modeling work (e.g., [11,12,13,14,15]) has been carried out in Northern Ireland, but long-term measured data does not currently exist. The European Commission Land Resource Management Unit has collected data on soil erosion throughout Europe [67], but the UK is not one of the contributing countries. Data from the Republic of Ireland could be used as an approximate analogue for Northern Ireland, but long-term field-based data at a local field-scale is required to more precisely determine the extent of the problem in the present day. It is postulated that only when soil erosion and its impacts are recognised as a problem by farmers as well as policy makers, and considerable long-term monitoring of erosion has been conducted, will consideration be given to the types of changes in policy outlined in this paper.

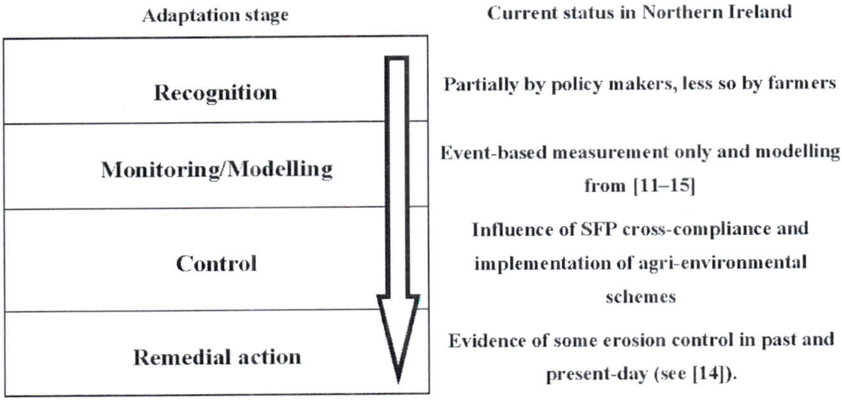


**Figure 3.** Stages required to achieve necessary adaptation measures for controlling soil erosion now and in the future and their current status in Northern Ireland (*after* [24]).

## 4. CONCLUSIONS

The off-site impacts of soil erosion in terms of sedimentation and nutrient enrichment of surface waters are a present-day problem in Northern Ireland, and recent studies have projected how these issues may become more serious under the impacts of future climate change. This review paper has examined the policies in place to manage erosion in Northern Ireland, and has sought to evaluate their effectiveness, before offering some suggestions on how policies may be reshaped to better manage an escalating problem. Soil erosion and its impacts are currently managed in an indirect way by the Common Agricultural Policy and a number of associated agri-environmental schemes, yet farmers are often not keeping their land in good agricultural and environmental condition, leading particularly to off-site impacts of erosion which are affecting water quality. Suggestions for policy changes were then offered, including a range of measures at a range of spatial scales. In terms of changes at the EU level, such as reforms to the WFD and the CAP, the use of clearer environmental standards for water quality targets may provide a more rigorous basis for ensuring thresholds are not exceeded. At a governmental level, ideas include greater enforcement of subsidy requirements, educating farmers on soil erosion prevention and remediation, implementation of conservation measures in high-risk zones as part of the SFP cross-compliance, and consideration to the development of voluntary agri-environmental schemes aimed directly at reducing soil erosion. The need for long-term measured erosion data to help raise awareness amongst farmers and policy makers, as well as help validate models to assess the future problem is a pressing research requirement to help ensure such measures can be taken in the future to manage a problem which looks set to worsen over the coming decades.

## ACKNOWLEDGMENTS

As this research was largely undertaken as part of my Ph.D., I wish to acknowledge my supervisors Dave Favis-Mortlock (University of Oxford) and Rowan Fealy (NUI Maynooth) for their help with the future modelled data which lies behind some of the ideas examined in this paper. I am also grateful to The Department of Employment and Learning Northern Ireland for funding my Ph.D.

## REFERENCES

1. *Soil Quality and Soil Erosion*; Lal, R., Ed.; Soil and Water Conservation Society: Ankeny, IA, USA, 1999.

2. Morgan, R.P.C. *Soil Erosion and Conservation*, 3rd ed.; Blackwell Publishing Limited: Oxford, UK, 2005.

3. He, X.; Zhou, J.; Zhang, X.; Tang, K. Soil erosion response to climatic change and human activity during the Quaternary on the Loess Plateau, China. *Reg. Environ. Chang.* **2006**, *6*, 62–70.

4. Lal, R.; Stewart, B.A. *Soil Degradation*; Springer-Verlag: New York, NY, USA, 1990.

5. Pimentel, D. Soil erosion: A food and environmental threat. *Environ. Dev. Sustain.* **2006**, *8*, 119–137.

6. Langdale, G.W.; West, L.T.; Bruce, R.R.; Miller, W.P.; Thomas, A.W. Restoration of eroded soil with conservation tillage. *Soil Technol.* **1992**, *5*, 81–90.

7. Pimentel, D.; Harvey, C.; Resosudarmo, P.; Sinclair, K.; Kurz, D.; MnCair, M.; Crist, S.; Shpritz, L.; Fitton, L.; Saffouri, R.; *et al*. Environmental and economic costs of soil erosion and conservation benefits. *Science* **1995**, *267*, 1117–1123.

8. Wardle, D.A.; Bardgett, R.D.; Klironomos, J.N.; Setälä, H.; van der Putten, W.H.; Wall, D.H. Ecological linkages between aboveground and belowground biota. *Science* **2004**, *304*, 1634–1637.

9. Boardman, J. A short history of muddy floods. *Land Degrad. Dev.* **2010**, *21*, 303–309.

10. Boardman, J.; Shepheard, M.L.; Walker, E.; Foster, I.D.L. Soil erosion and risk-assessment for on- and off-farm impacts: A test case using the Midhurst area, West Sussex, UK. *J. Environ. Manag.* **2009**.

11. Mullan, D.J.; Favis-Mortlock, D.T.; Fealy, R. Addressing key limitations associated with modelling soil erosion under the impacts of future climate change. *Agric. For. Meteorol.* **2012**, *156*, 18–30.

12. Mullan, D.J.; Favis-Mortlock, D.T. Managing soil erosion: A case study from Ireland. *Geogr. Rev.* **2011**, *24*, 24–26.

13. Favis-Mortlock, D.T.; Mullan, D.J. Soil Erosion by Water under Future Climate Change. In *Soil Hydrology, Land Use and Agriculture: Measurement and Modelling*; Shukla, M., Ed.; CABI Publishing: Wallingford, UK, 2011; pp. 384–414.

14. Mullan, D.J. Soil erosion on agricultural land in the north of Ireland: Past, present and future potential. *Ir. Geogr.* **2013**.

15. Mullan, D.J. Soil erosion under the impacts of future climate change: Assessing the statistical significance of future changes and the potential on-site and off-site problems. *Catena* **2013**, *109*, 234–246.

16. Favis-Mortlock, D.T.; Savabi, M.R. Shifts in Rates and Spatial Distributions of Soil Erosion and Deposition under Climate Change. In *Advances in Hillslope Processes*; Anderson, M.G., Brooks, S.M., Eds.; Wiley: Chichester, UK, 1996; Volume 1.

17. Williams, J.; Nearing, M.A.; Nicks, A.; Skidmore, E.; Valentine, C.; King, K.; Savabi, R. Using soil erosion models for global change studies. *J. Soil Water Conserv.* **1996**, *51*, 381–385.

18. Favis-Mortlock, D.T.; Guerra, A.J.T. The implications of general circulation model estimates of rainfall for future erosion: A case study from Brazil. *Catena* **1999**, *37*, 329–354.

19. Nearing, M.A. Potential changes in rainfall erosivity in the U.S. with climate change during the 21st century. *J. Soil Water Conserv.* **2001**, *56*, 229–232.

20. Pruski, F.F.; Nearing, M.A. Climate-induced changes in erosion during the 21st century for eight U.S. locations. *Water Resour. Res.* **2002**, *38*, Article 1298.

21. Trenberth, K.E.; Dai, A.; Rasmussen, R.M.; Parsons, D.B. The changing character of precipitation. *Bull. Am. Meteorol. Soc.* **2003**, *84*, 1205–1217.

22. Nearing, M.A.; Jetten, V.; Baffaut, C.; Cerdan, O.; Couturier, A.; Hernandez, M.; Le Bissonnais, Y.; Nichols, M.H.; Nunes, J.P.; Renschler, C.S.; *et al.* Modeling response of soil erosion and runoff to changes in precipitation and cover.*Catena* **2005**, *61*, 131–154.

23. Rosenzweig, C.; Hillel, D. *Climate Change and the Global Harvest. Potential Impacts of the Greenhouse Effect on Agriculture*; Oxford University Press Inc.: New York, NY, USA, 1998.

24. Verspecht, A.; Vandermeulen, V.; de Bolle, S.; Moeskops, B.; Vermang, J.; van den Bossche, A.; van Huylenbroeck, G.; de Neve, S. Integrated policy approach to mitigate soil erosion in West Flanders. *Land Degrad. Dev.* **2011**, *22*, 84–96.

25. Prazan, J.; Dumbrovsky, M. Soil conservation policies: Conditions for their effectiveness in the Czech Republic. *Land Degrad. Dev.* **2010**, *21*, 124–133.

26. Schuler, J.; Sattler, C. The estimation of agricultural policy effects on soil erosion-An application for the bio-economic model MODAM, 2010. *Land Use Policy* **2010**, *27*, 61–69.

27. Lundekvam, H.E.; Romstad, E.; Øygarden, L. Agricultural policies in Norway and effects on soil erosion. *Environ. Sci. Policy* **2003**, *6*, 57–67.

28. Boellstorff, D.; Benito, G. Impacts of set-aside policy on the risk of soil erosion in central Spain. *Agric. Ecosyst. Environ.***2005**, *107*, 231–243.

29. Evans, R. Runoff and soil erosion in arable Britain: Changes in perception and policy since 1945. *Environ. Sci. Policy* **2010**, *13*, 141–149.

30. DEFRA. *Piloting a Cost-Effective Framework for Monitoring Soil Erosion in England and Wales*; SP1311; Department for Environment Food and Rural Affairs: London, UK, 2013.

31. European Commission. *Thematic Strategy for Soil Protection (Communication)*; Brussels, 22-9-2006. COM (2006) 231 final; European Commission: Brussels, Belgium, 2006.

32. Jones, A.; Panagos, P.; Barcelo, S.; Bouraoui, F.; Bosco, C.; Dewitte, O.; Gardi, C.; Erhard, M.; Hervás, J.; Hiederer, R.; *et al. The State of Soil in Europe*; European Commission Joint Research Centre: Ispra, Italy, 2012.

33. European Commission. *Common Agricultural Policy. A Partnership between Europe and Farmers*; Publications Office of the European Union: Luxembourg, Luxembourg, 2012; p. 16.

34. DEFRA. *Soil Protection Review 2010 GAEC 1*; Department for Environment Food and Rural Affairs: London, UK, 2010.

35. European Commission. *Establishing a Framework for Community Action in the Field of Water Policy*; Directive 2000/60/EC of the European Parliament and of the Council; European Commission: Brussels, Belgium, 2000.

36. Lilly, A.; Grieve, I.C.; Jordan, C.; Baggaley, N.J.; Birnie, R.V.; Futter, M.N.; Higgins, A.; Hough, R.; Jones, M.; Noland, A.J.; *et al. Climate Change, Land Management and Erosion in the Organic and Organo-Mineral Soils in Scotland and Northern Ireland*; Scottish Natural Heritage Commissioned Report No. 325 (ROAME No. F06AC104—SNIFFER UKCC21); Scottish Natural Heritage: Edinburgh, Scotland, UK, 2009.

37. DEFRA. Common Agricultural Policy Reform, (On-Line), Department for the Environment, Food and Rural Affairs, London, 2011. Available online: http://www.defra.gov.uk/food-farm/farm-manage/cap-reform/ (accessed on 12 October 2011).

38. Royal Commission on Environmental Pollution (RCEP). *Sustainable Use of Soil*; 19th Report; Her Majesty's Stationery Office (HMSO): London, UK, 1996.

39. Silvey, V. The contribution of new varieties to cereal yields in England and Wales between 1947 and 1983. *J. Natl. Inst. Agric. Bot.* **1986**, *17*, 155–168.

40. Baird, W.W.; Tarrant, J.R. *Hedgerow Destruction in Norfolk 1946–1970*; School of Environmental Sciences, University of East Anglia: Norwich, UK, 1973.

41. Environment Agency. *The State of the Environment of England and Wales: The Land*; Environment Agency, Her Majesty's Stationery Office: London, UK, 2000.

42. Evans, R.; Cook, S. Soil erosion in Britain. *SEESOIL* **1986**, *3*, 28–59.

43. Favis-Mortlock, D.T. Ireland. In *Soil Erosion in Europe*; Boardman, J., Poesen, J., Eds.; Wiley: Chichester, UK, 2006.

44. Robinson, D.A. Agricultural practice, climate change and the soil erosion hazard in parts of southeast England. *Appl. Geogr.* **1999**, *19*, 13–27.

45. Boardman, J. Soil erosion and flooding on the eastern South Downs, southern England, 1976–2001. *Trans. Inst. Br. Geogr.* **2003**, *28*, 176–196.

46. Boardman, J.; Evans, R.; Ford, J. Muddy floods on the South Downs, southern England: Problem and responses.*Environ. Sci. Policy* **2003**, *6*, 69–83.

47. Evans, R.; Boardman, J. Curtailment of muddy floods in the Sompting catchment, West Sussex, southern England. *Soil Use Manag.* **2003**, *19*, 223–231.

48. Evans, R. Curtailing water erosion of cultivated land: An example from north Norfolk, eastern England. *Earth Surf. Processes Landf.* **2006**, *31*, 598–605.

49. European Commission. *On Support for Rural Development from the European Agricultural Guidance and Guarantee Fund*; Council Regulation (EC) No. 1257/1999; European Commission: Brussels, Belgium, 1999.

50. Department of Agriculture and Rural Development (DARD). *Farm Nutrient Management Scheme (Northern Ireland) 2005 Scheme Booklet FNMS I*; Department of Agriculture and Rural Development: Belfast, UK, 2005; p. 50.

51. Fealy, R.; Sweeney, J. Statistical downscaling of temperature, radiation and potential evapotranspiration to produce a multiple GCM ensemble mean for a selection of sites in Ireland. *Ir. Geogr.* **2008**, *41*, 1–27.

52. Mullan, D.J.; Fealy, R.; Favis-Mortlock, D.T. Developing site-specific future temperature scenarios for Northern Ireland: Addressing key issues employing a statistical downscaling approach. *Int. J. Climatol.* **2012**, *32*, 2007–2019.

53. Lempert, R.; Nakicenovic, N.; Sarewitz, D.; Schlesinger, M. Characterizing climate change uncertainties for decision-makers - an editorial essay. *Clim. Chang.* **2004**, *65*, 1–9.

54. European Commission. *On the Quality of Fresh Waters Needing Protection or Improvement in Order to Support Fish Life*; Directive 2006/44/EC of the European Parliament and of the Council; European Commission: Brussels, Belgium, 2006.

55. Collins, A.L.; Anthony, S.G. Assessing the likelihood of catchments across England and Wales meeting "good ecological status" due to sediment contributions from agricultural sources. *Environ. Sci. Policy* **2008**, *11*, 163–170.

56. Verheijen, F.G.A.; Jones, R.J.A.; Rickson, R.J.; Smith, C.J. Tolerable *versus* actual soil erosion rates in Europe. *Earth Sci. Rev.* **2009**, *94*, 23–38.

57. DARD. *Guide to the Single Farm Payment (SFP) Scheme*; Department of Agriculture and Rural Development: Belfast, UK, 2011; p. 39.

58. MAFF. *Soil Erosion by Water. ADAS Leaflet 890*; Ministry of Agriculture, Fisheries and Food: Alnwick, UK, 1985.

59. DEFRA. *Controlling Soil Erosion: A Manual for the Assessment and Management of Agricultural Land at Risk of Water Erosion in Lowland England*; Revised September 2005; Department for the Environment, Food and Rural Affairs: London, UK, 2005.

60. Rural Payments Agency (RPA). *Single Payment Scheme: Cross-Compliance Guidance for Soil Management 2010 Edition*; Rural Payments Agency/Department for Environment Food and Rural Affairs: London, UK, 2010; p. 76.

61. Evans, R. Outdoor pigs and flooding: an English case study. *Soil Use Manag.* **2004**, *20*, 178–181.

62. Verstraten, G.; Poesen, J.; Govers, G.; Gillijns, K.; van Rompaey, A.; van Oost, K. Integrating science, policy and farmers to reduce soil loss and sediment delivery in Flanders, Belgium. *Environ. Sci. Policy* **2003**, *6*, 95–103.

63. Vandaele, K.; Lammens, J.; Priemen, P.; Evrard, E. How to control muddy floods from cultivated catchments? Lessons from the Melsterbeek catchment in Flanders (Belgium), (On-line), Samenkering Land en water, St-Truiden, Belgium, 2013. Available online: http://www.land-en-water.be/english.html (accessed on 3 March 2013).

64. DARD. *Countryside Management Scheme (Northern Ireland)*; Scheme Information Booklet. Department of Agriculture and Rural Development: Belfast, UK, 2007; p. 144.

65. Boardman, J.; Vandaele, K. Soil erosion, muddy floods and the need for institutional memory. *Area* **2010**, *42*, 502–513.

66. Louwagie, G.; Gay, H.S.; Sammeth, F.; Ratinger, T. The potential of European Union Policies to address soil degradation in agriculture. *Land Degrad. Dev.* **2009**, *22*, 5–17.

67. Panagos, P.; van Liedekerke, M.; Jones, A.; Montanarella, L. European Soil Data Centre (ESDAC): Response to European policy support and public data requirements. *Land Use Policy* **2012**, *29*, 329–338.

# Chapter 12

# Optimization of Soil Erosion and Flood Control Systems in the Process of Land Consolidation

**Miroslav Dumbrovsky[1] and Svatopluk Korsuň[1]**

[1] Brno University of Technology, Faculty of Civil Engineering, Department of Landscape Water Management, Czech Republic

## 1. INTRODUCTION

Extreme hydrological phenomena of recent years have highlighted a well-known fact that it is necessary to pay greater attention to the problems of flood-prevention and soil erosion control on a large part of the Czech Republic. Case study areas are the most endangered territories. The case study area was selected as a case study mainly for its natural conditions and high risk of soil degradation and occurrence of flash floods. Relief, geomorphology, the present state of the complex system of soil properties, the types of agricultural farming practices and land use, are all contributing to accelerated soil erosion and runoff with all its negative impacts on the built-up areas.

The main soil degradation problems in the case study area are soil erosion caused by water, soil compaction and decline in organic matter. Soil erosion is fostered by i) soil degrading (intensive) farming practices such as up and down hill conventional tillage and other conventional agricultural operations on arable land, ii) frequent extreme hydrological events, and iii) a decreasing ability of soils for water retention (decline in organic matter and land conversion). Soil compaction is a problem due to intensive conventional farming on arable land (using heavy machinery). The decline in organic matter results from the constant soil erosion process. Main causes of decline in organic matter are conventional farming practices without using manure and other organic matter. Decline in organic matter causes a decrease of natural crop productivity of soil and decreases yield.

Great runoffs occur on these areas and transform into flood waves in watercourses. Forest grounds are also affected, especially in case of unsuitable transport, wood cut and growth make-up. Solving of the problems of territory protection from unfavourable and damaging effects of overland water flow must therefore begin in catchments areas and particularly during any interference with landscape. Appropriate conservation measures are required to prevent and reduce runoff and soil degradation resulting from intensive agriculture. The adoption of the most appropriate practices and optimisation the farming conservation system it is necessary to carry out analyses and evaluations of the erosion rate and the basic characteristics of runoff in given sub-catchments. This system of evaluation provides information about erosion and runoff risks plots and serves for decision making regarding soil conservation and flood prevention measures. The success of the system of soil conservation depends on suitable technical assistance and support from responsible state organisations (Ministry of Agriculture and Ministry of Environment), sufficient sources of information as well as the ability and willingness of land users to adopt soil conservation measures. The main motivation for farmers to apply soil conservation measures is the economic motivation through financial subsidies along with penalties for farmers if they fail to comply with the rules of the funding program.

Nevertheless, when introducing a soil erosion control and flood prevention measures in a certain watershed, best management practices are mostly to be able decrease of erosion rate but unable to restrict a surface runoff substantially. For that reason it is necessary to apply a whole system of soil conservation measures. In places with long slopes technical and biotechnical soil erosion control practices (primarily of linear character) are necessary. These technical measures are broad base terraces and channels in case study area. These biotechnical measures together with the implementation of grassed courses of concentrated surface runoff (grassed waterways) create an appropriate network of new hydrolines in the watershed. Biotechnical line elements of soil erosion control serve as permanent barriers or obstacles for water runoff and are designed in order to determine, by their location, the ways of land management. Some technical and biotechnical measures could be suitable regarding their technical feasibility, economic efficiency and environmental effectiveness. The spatially and functionally limited soil-conservation system in a given teritorry offers spaces and lines in which it would be possible to locate territorial systems of ecological stability under certain conditions. Soil conservation and flood prevention practices, connected with territorial systems of ecological stability can be characterized as desirable anthropogenic landscape-forming elements. These would form the appearance of the landscape and significantly enhance natural processes in the region. They create suitable biological conditions in spite of the fact that they mostly do not meet qualitative and dimensional characteristics of biocentres and biocorridors.

Highly fragmented land ownership is prevalent in the area Biotechnical and technical soil conservation measures cannot be applied without respecting property rights. Integral parts of any project of soil erosion control (its basic network) are usually line elements for soil erosion control (broad base terraces and channels etc.), which run across individual owners' fields. Therefore it is necessary to identify every owner and discuss with the project and relevant proposals. The greatest interventions with agricultural landscape are land consolidation which, apart from other less important objectives, are designed to completely eliminate or at least partly limit unfavourable effects of runoff (especially soil erosion) and thus to become one of the most important elements of territory organisation and protection. Therefore it was found suitable to design the system of the soil and water conservation in the process of land consolidation in the Czech Republic.

Optimal spatial and functional delimitation of soil erosion control practices in the landscape is one of the basic steps in the plan of comprehenshive land consolidation, in addition to the implementation of a new network of field roads and landscape features enhancing ecological stability. Soil erosion control and flood prevention practices are included in the system of public facilities within the framework of the land consolidation process (where property relations are consistently solved).

The definition of land consolidation is from Act No. 139/2002 Coll., on Reparcelling and Land Authorities and amending Act No. 229/1991 Coll., on the Arrangement of Ownership titles to Land and other Agricultural Assets, as amended. The land consolidation processes in case study area have started in 2005.

This procedure gives solutions to the whole area, both from the aspect of a new land and ownership arrangement (Figure 2) and from the aspect of soil conservation and flood prevention and improvement of environment (Figure 1).

**Figure 1.** Soil conservation and Flood prevention system.

**Figure 2.** Parcel of owners before and after land consolidation.

Recently, the process of comp ex land consolidation in the Czech Republic has provided a unique opportunity for improving the quality of the environment and sustainability of crop production through better soil and water conservation. The current process of the land consolidation consists of the rearrangement of plots within a given territory, aimed at establishing the integrated land-use economic units, consistent with the needs of individual land owners and land users.

Integrated territory protection can be reached by controlling runoff by means of design of terraces as a soil erosion control measures. A number of mathematical models, mostly simulation ones, to solve water-management problems have been compiled, some of which include the option of exact mathematical optimization. A certain summary of these models, including their characteristics and application possibilities, were elaborated by Kos (1992). An interesting combination of the application of a simulation and optimization model technique in the elaboration of design of a particular water-management system was described by Major, Lenton et al. (1979), a three-model approach to solve water-management systems was used by Onta, Gupta and Harboe (1991). Benedini (1988) dealt more generally with the design and possible applications of these models. Most likely, an optimization model has not been designed, which would enable to attach territory protection and the measures to eliminate the amount and accumulation of runoff in catchments areas to solving water-management problems.

The created procedure is a universal tool which can be applied for any territory. It enables to find the most suitable combination of all possible alternatives of various erosion controls and flood protection measures under given conditions of each particular site. Such sites do not always have to be ground used for farming. They may also include in forest or urban areas or site arrays in various territories.

## 2. METHOD

The optimization process of designing the system of *integrated territory protection* (the *IOU* system) begins with the processing of the system of organisational, agrotechnical, biotechnical and technical measures at individual sites of the case study territory. It is necessary to derive hydrograms of direct runoff from extreme rainfall events for each of these variants. Then it is necessary to elaborate the variants of terraces and other conservation measures on all sections of watercourses and variant of designs of retention protection reservoirs. Not only rivers, streams and brooks are included into the watercourse category within this procedure but also sometimes passed watercourses such as terraces, grass infiltration belts or the lines of stabilisation of concentrated runoff waterways in valley lines.

A selection of the most suitable combination of all prepared variants is listed. With respect to the fact that it is necessary to find optimal dimensions for some of the system elements, there is usually a great number, in case of a continual solving even an infinite number, of possible combinations. It is therefore necessary to use an optimalized mathematical model to find the most suitable combination. This model was created on the basis of a mixed discrete programming (Korsuň et al., 2002, Dumbrovský et al., 2006). Its basic building stones are three generally formulated partial models: *A.* partial model of protective measures at individual sites of the case study area. *B.* partial model of a watercourse. *C.*partial model of a reservoir.

It is possible to shape an *optimization model of integrated territory protection* (*OMIOU*) from these partial models for any particular territory. The partial models are repeatedly inserted into the *OMIOU*as needed so as to exactly copy the modelled system structure. It is necessary to determine in advance one criterion or more simultaneously operating optimization criteria for each optimization function. A whole range of criteria can be determined for a given purpose. These can be taken from the sphere of economy but also from those of ecology, water-management, social etc. However it is necessary to define the most suitable criteria as far as quality is concerned but also to have a chance to quantify the values of each defined criterion. On top of that, it is necessary, in case of several simultaneously operating optimization criteria, to assign each criterion its adequate weight with which it will enter the solving process and which will support its effect on the result, so called a compromise solution in competition with the other criteria.

In creating the procedure of the *IOU* system proposal optimization in connection with the process of territory organisation a requirement of a maximal protection of inhabited and other areas with the exertion of minimal means was formulated for the solving process on the level of land consolidation as one of the suitable optimization criteria. It is a criterion consisting of three simultaneously operating partial economic, but at the same time water-management and socially aimed at their impacts. Criteria include:

- minimization of the average annual damage (material damage: it is estimated that input requirements and conditions will not allow solutions which could lead to losses of human lives) originated by overland runoffs from rainfall events and then by their concentration in watercourses.

- minimization of the average annual economic losses in farming production related to the realisation of proposed protective measures on arable land.

- Minimization of the average annual expenses (the sum of expenses for running and maintenance plus the amortization of the capital goods) of the proposed conservation measures.

Seeing that in most cases they are average annual values, quantified for example in thousands of CZK per year, these criteria can be assigned the same weights 1:1:1 in reflection.

The optimization mathematical model is a system of equations, which model a given system behaviour, the variables in the equation describe a system structure and the dimensions of its individual elements. Non-equations found in each model are transformed into equations by means of additional variables in the course of the model solving process, therefore the term *equation* is used only. The above mentioned partial models were created in the modelling and calculation system *GAMS* (*General Algebraic Modelling System*) in its general form (Charamza, 1993) so it can be used to model any integrated territory protection system. The nature of the solved problems implies that the defining process of all the variables used in the model as positive variables. They can be either continuous ones which are marked $x$ herein after or binary ones (they can take on only 0 or 1 values) marked with the symbol $x\,B$. Other symbols are used to mark variables and coefficients. Activities proceeding in time must be modelled in the whole system according uniform timekeeping.

*The partial model A* is aimed at terraces and other biotechnical, agrotechnical, and organisation conservation measures in the catchments area of a certain watercourse. These measures are usually designed within land consolidation to decrease overland flow of rainfall events and thus to limit the effects of soil erosion and damage in inhabited territories. The various proposals of protective measures must be elaborated in each individual case before an optimization model is designed (pre-optimization) as pragmatically created systems of various, mutually complementary interventions with the individual catchments area elements. Such a partial catchments area element could be, for example, valley and slope area above one bank of a certain watercourse section in the range from the bank line to the interstream divide line.

The part of runoff from the design rainfall events which will not be caught by the system of catchments area protective measures (*residual runoff*) will concentrate in a particular watercourse and will create a design Q runoff or flood wave. The time $T$ of passage of the design flood wave through a watercourse will be divided into $r$ of equally long *time intervals* (*TI*); time $t$ of the durance of one *TI* will thus be given by the relation $t = T / r$. For the individual *TI*s, partial volumes $w\ 1$ of the design flood wave are then quantified, $i = 1, 2,..., r$.

In case of the application of the above mentioned optimization criterion, the following indicators must be quantified for each pre-optimization processed variant of the protective measure set on a partial catchments area element:

- its estimated effect $U$ expressed financially as an average annual level of damage on land, growth, buildings, roads etc. which will occur after the variant has been realised (*residual damage*),

- estimated average annual economical loss $E$ in farming production related to the realisation of the proposed measures on arable land.

- realisation costs of a particular variant and its average annual own costs $N$,

- the amount of residual runoff $Oi$ into a watercourse in the individual *TI*s.

These data represent input information for the partial model *A*. In the course of the optimization process, only one – optimal – variant with the most suitable indicators will be chosen from thus prepared variants of systems of protective measures for each partial catchments area element. Residual runoffs concentrating in a watercourse runway from the watercourse adjacent partial areas protected by optimal systems of measures will cause a gradual accretion of a flood wave passing through the watercourse. The protection from damage which could be caused by this flood wave will be provided by the protective measures on the watercourse and retention protective reservoirs as mentioned later (the partial models *B* and *C*).

Binary variables can be used for modelling of individual variants of protective measure systems in each of the partial catchments area elements in a discrete way. The total number of catchments area elements will be *m*. If, for example, *n* variants of protective measure systems of a $d^{th}$ catchments area element are modelled by relations to binary variables $x_{B1dp} \in \{0, 1\}$, $d = 1, 2,..., m$, $p = 1, 2,..., n$, the effects of these measure systems for this catchments area element can be write into the model using the following equations:

the equation of protective effects (residual damage)

$$x_{Ud} = \sum_{p} U_{dp} \cdot x_{B1dp} \tag{1}$$

the equation of economic damage

$$x_{Ed} = \sum_{p} E_{dp} \cdot x_{B1dp} \tag{2}$$

the equation of own costs

$$x_{Nd} = \sum_{p} N_{dp} \cdot x_{B1dp} \tag{3}$$

the equation of residual runoff, i.e. contribution of a $d^{th}$ catchments area element to the flood wave volume on a particular watercourse section in $i^{th}$ *TI*

$$x_{Oid} = \sum_{p} O_{idp} \cdot x_{B1dp} \tag{4}$$

for $i = 1, 2,..., r$,

$d = 1, 2,..., m$,

$p = 1, 2,..., n$,

where $x_{Ud}$ is the total residual damage in a $d^{th}$ catchments area element,

$x_{E,Ud}$ is the total economic loss in a $d^{th}$ catchments area element,

$x_{Oid}$ is the total residual runoff from a $d^{th}$ catchments area element in $i^{th}$ *TI*.

Because only one of the protective measure system variants can enter the solving process, the following condition must be valid for the sum of all the binary variables of a $d^{th}$ catchments area element:

$$\sum_{p} x_{B1dp} = 1 \tag{5}$$

*The partial model B* captures the passage of the design flood wave through the watercourse sections. The sections are either left in their present state, the optimization of a river bed or a contour furrow systems design (including the building of protective dams), or the reconstruction an earlier carried out adjustment or protective dams may be required. A watercourse section can also be a water or dry protective reservoir which will be modelled in a way mentioned in the partial model C description.

Flood damage that can occur is quantified for each watercourse section during its modelling. Further, runoffs from the section are calculated in the individual *TI*s of a flood wave passage. With respect to the overland flow from the initial section profile to the last one, it is necessary to determine a time shift which will affect collisions of flood waves on the main watercourse and at the mouths of its tributaries. The mean value of the runoff volume which can be found in a section (in a river bed or also in an inundation territory) in the course of $i^{th}$ *TI* is at the position of the basic section variable. The values of the other variables are related to this variable: the variables of the water flowing through the section, time of concentration, the level of flood damage in the section, and the level of runoff from the section. The courses of these non-linear functions are derived from the watercourse pre-optimization variant designs. They are replaced with linear function part by part in the optimization model. The formulation of particular equations is mentioned in Chapter 5.1 of Patera, Korsuň et al. (2002).

*The partial model C* is outlined for a designed multipurpose water reservoir with unknown capacities of spaces protective controllable $x_{OO}$, protective non-controllable $x_{ON}$ and total $x_V$. The necessary volumes of the spaces of dead storage $S \geq 0$ and active storage capacity $Z \geq 0$ are constant – these values result from other than protective requirements. The objective of analysis is to find its dimension which, respecting the requirement to create the spaces *S* and *Z,* with its protective spaces will ensure the reduction of culminated runoff from the reservoir to its optimal level during the passage of the design flood wave. In cases when the designed water reservoir has only a protective function, the value of the *Z* variable is zero; the values of both variables are zero $S = Z = 0$ for a dry protective reservoir.

The unknown volume of the total reservoir space is a variable, whose value which is limited from above by the maximal value *V max* corresponding with the biggest realisable variant of the reservoir design during the pre-optimization solutions. From below it is limited by the minimal variant, still acceptable for practice, with the total volume *V min*.

We cannot forget a situation when building a reservoir will not be acceptable due to the used optimization criteria. It is therefore necessary to introduce a binary variable *x B2* ϵ *{0, 1}* into the set of variable values. If this variable has a zero value, the reservoir will not enter the solving process, if *x B2= 1*, the entry of the reservoir into solving is cleared. Then the volume of the total reservoir space (without evaporation and percolation) must correspond with the following conditions

$$x_V = (S + Z) \cdot x_{B2} + x_{OO} + x_{ON} \tag{6}$$

$$V_{min} \cdot x_{B2} \leq x_V \leq V_m \tag{7}$$

The equations modelling the passage of the design flood wave through a dam profile, the calculations of the volumes of individual reservoir spaces and of necessary financial means are described in Chapter 5.1 of Patera, Korsuň et al. (2002). The partial model C can be also used for already an existing reservoir with a constant volume of the total space.

The model compilation from the fore mentioned partial elements in the presented form requires the introduction of a set of concrete coefficients and variables into the model for the model equation system to copy completely a particular system of *IOU*. These coefficients and variables should be derived from the pre-optimization processed background materials. In the case of non-standard requirements of an *IOU* system structure, it is necessary to introduce other equations to the model. Such new equations would capture these requirements. The model solving process in carried out on a computer by means of some of the *GAMS* system tools.

# 3. MATERIAL

To verify the function and potential of the already described optimization procedure, a system of integrated territory protection was chosen that was proposed within the framework of land consolidation on the case study area between the town of Hustopeče and the village of Starovice in the Czech Republic (see Figure 1). The declining ground in this region is mostly used as arable land. Overland flow is concentrated into its main waterway, which enters the residential parts of Hustopeče. Considerable, and frequently repeating damage, is caused by soil erosion on farm crops, sediment transport from arable land and especially by flooding parts of the town.

The proposed system of integrated protection of this farming territory and town is based on a system of technical-biotechnical, organisational and agrotechnical soil erosion control measures on arable land and of two conservation measures: 1. transfer of concentrated runoff from the drainage furrow or channel *K1* in the main valley line over the terrain into the adjacent valley line and creating a channel*K2* entering watercourse, 2. building a dry protective reservoir (polder) *P1* to catch parts of runoffs from the main valley line and another polder on the channel *K2* in the adjacent valley line above the village of Starovice.

The *IOU* system design for the given territory is based on the situation which would occur during a rainstorm with hundred-year periodicity (design rainfall). The protective measures with pre-optimization were designed in ten different variants, volume and cost (own costs) functions were derived for both the polders. It is estimated that *P1* polder filling, which is a side basin for the channel *K1*, will proceed through the channel side overfall. For the individual soil erosion control measure alternatives volumes and accumulation of overland runoffs, derived from the design rainfall, in the form of runoff hydrograms from two catchments areas: from the polder *P1* catchments and from the polder *P2* catchments. The passage of runoff waves through dam profiles of both polders takes from 510 minutes in the alternatives 1 and 2 to 195 minutes in the alternatives 9 and 10. It requires limiting the culmination water passages in the river beds below the two polders: below the *P1* this passage (runoff from the *P1*).

which will enter the city sewerage system in Hustopeče, should not exceed 0.125 $m^3.s^{-1}$, below the *P2* the passage limit should be, with regard to the protection of Starovice, chosen at 1.5 $m^3.s^{-1}$ at most and in variants of 1.0 and 0.5 $m^3.s^{-1}$ to determine the effect of this passage size on the *IOU* system optimal solution.

The optimization model consists of 3,506 equations with the total of 1,673 structural variables, 539 of which are binary variables. The model objective function (optimization criterion) minimises the sum of average annual values of flood damages, economic losses and biotechnical measures and polders own expenses in the proportion of 1:1:1. It is ensured that only one protection system alternative can enter the *OMIOU* optimal solution in both the catchments areas, but it can be different for each of the catchments. These alternatives are marked as *A1* and a particular alternative number for the *P1* polder catchments, the *P2* polder was allocated symbol *2* in a similar way. The polders can enter the solution but they also do not have to. The runoff wave from the *P1* polder catchments may be partly or completely transferred into the *P2* polder. Permissible maximum of water depth in the *P1* polder is 5.0 m, it is 4.34 m in the *P2* polder. The *P1* polder low outlet dimensions (the inside diameter of outlet pipeline of a round shape) *d* = 200 mm, there is a possibility of choice from *d* = 200, 300, 400, 500 or 600 mm for the *P2* polder.

## 4. RESULTS AND DISCUSSION

The model function and behaviour were examined first in relation to the project research objectives. Then the possibilities of experimentation on the model of the designed system were tested (Korsuň et al., 2002). The optimization process was carried out with the three above listed values of admissible maximal runoff from the *P2* polder and then in an experimental way with various runoffs from both catchments areas: with real runoffs derived from hundred-year rain storms for the individual variants of conservation measures in both the catchments areas, and with fictive multiples of these runoffs.. Variants with other changes in input conditions (e.g. without the polders entering the solving process) were calculated for the same reason. The results of these solutions are not listed here. Optimal solutions of variants No. 1, 3 and 5 correspond with the real state of the input conditions. These solution results were derived from overland flows from the grounds and from three real values of admissible maximal runoff from the *P2* polder above Starovice. The results of following experiments on the optimization model have led to a number of interesting findings. However, the most important finding is the fact that the experimental locality can be protected as required without any interference into plant production conditions, i.e. without any (on site) economic loss on the produce only by conservation measures themselves: by draining overland and hypodermic runoffs through contour furrows and channels in the *P2* polder. This protective system design is valid only provided the applied optimization criterion is kept. The resulting design can be different in the case of any change to the criterion (e.g. the changes in the weights of the three used partial criteria) or in case of the application of a different criterion.

## 5. CONCLUSION

The results of the practical application of the optimization procedure in designing terraces and retention reservoirs within integrated territory protection verify its functionality and applicability. In cases when it is not clear in advance which of the potential torrential rainfall could be the most dangerous, the model will provide solutions with all chosen rainfalls types for the result to comply with the territory protection requirements.

The created model can be used to find either one optimal solution or, in case it is necessary to verify the position of the optimal solutions with the changes of some input conditions and requirements, more times in more versions with variables and coefficients modified by these changes. The possibility of multiple application of this model and to obtain a whole set of optimal solutions visualises much better the character and behaviour of the designed system in reactions to modifications of the input conditions and requirements and thus enables to improve significantly the process of making decisions about the design final shape.

A great advantage of the model lies in the general formulation of its components – partial models of conservation measures at individual sites of the experimental locality, watercourse and reservoir. This should enable its problem-free application for optimization design of integrated territory protection under any conditions and at any site

# REFERENCES


1. M. Benedini, 1988 Developments and possibilities of optimisation models Agric. Water Manag. 13 329 358

2. M. Dumbrovský, et al. 2003 Optimisation of the system of soil and water conservation for runoff minimizing in certain watershed in the process of land consolidation (in Czech) Final research report, NAZV-QC1292 VÚMOP, Praha

3. P. Charamza, et al. 1993 Modelling system GAMS (in Czech). MFF UK, Praha.

4. S. Korsuň, et al. 2002 Creating and verification of the model for optimisation of soil and water conservation (in Czech) Final research report, A01-NAZV-QC1292 FAST VUT, Brno

5. Z. Kos, 1992 Water management systems and their mathematical models during climate changes (in Czech) Vod. Hosp. 7 211 216

6. D. C. Major, R. L. Lenton, et al. 1979 Applied water resource systems planning. Prentice- Hall Inter., Inc., London.

7. P. R. Onta, A. D. Gupta, R. Harboe, 1991 Multistep planning model for conjunctive use of surface and groundwater resources Jour. Water Res. Plan. Manag. 6 662 678

8. A. Patera, J. Váška, J. Zezulák, V. Eliáš, S. Korsuň, et al. 2002 Floods: prognosis, water streams and landscape (in Czech) ČVUT / ČVVS Praha

# Index